OUTSIDE

the

GATES *of*

SCIENCE

OUTSIDE

the

GATES *of*

SCIENCE *

Damien Broderick

* *why it's time for the paranormal to come in from the cold*

Thunder's Mouth Press
New York

OUTSIDE THE GATES OF SCIENCE:
Why It's Time for the Paranormal to Come In from the Cold

Published by
Thunder's Mouth Press
An Imprint of Avalon Publishing Group, Inc.
245 West 17th Street, 11th Floor
New York, NY 10011
www.thundersmouth.com

AVALON
publishing group incorporated

Library of Congress Cataloging-in-Publication Data is available.

ISBN-13: 978-1-56025-986-2
ISBN-10: 1-56025-986-8

9 8 7 6 5 4 3 2 1

Interior book design by Bettina Wilhelm

Printed in the United States of America
Distributed by Publishers Group West

To Len Kane

[T]here are no truths outside the Gates of Eden.
                    —Bob Dylan, "The Gates of Eden"

Drive the Pseudos Out of the Workshop of Science
    —John Archibald Wheeler, essay title, *Skeptical Enquirer* 3, 1979

Decades of cumulating laboratory evidence strongly suggest that real correlations exist between mental states and randomized events that are distant in space and time, some even in the future. The laboratory findings are generally weaker than those reported from the field and because of the experimental paradigms used, they are basically statistical in nature. In other words they need a large number of trials to reach statistical significance. What is worse is that it is difficult to replicate the phenomena even when allowing for their intrinsic statistical character. Because of the claimed transcendental nature of the phenomena the experimenter is an intrinsic part of the experiment and therefore replication from one experimenter to another is not expected to be easily accomplished. Replication by independent scientists is of course the requirement for a phenomenon to be considered objective and real. Therefore the controversy about the reality of these phenomena continues in spite of the cumulating evidence . . . [Philosopher Daniel Dennett declared] that he would "commit suicide if paranormal phenomena turned out to be real."
        —Dick J. Bierman, "On the Nature of Anomalous Phenomena"

CONTENTS

∾

1.
THE SUN ALSO RISES

While critics are fond of relating, as Professor Hyman does in his report, that there has been "more than a century of parapsychological research . . . ," psychologist Sybo Schouten . . . has noted that the total human and financial resources devoted to parapsychology since 1882 is at best equivalent to the expenditures devoted to fewer than two months of research in conventional psychology in the United States.
—Dr. Jessica Utts, "Evaluation of Program on Anomalous Mental Phenomena"

You've heard the following kind of story before, and maybe something like it has happened to you or a family member, but you still don't know what to *do* with it, what to believe:

☙☙

July 6, 1996, was a sunny day in Texas. Barbara, the lawyer and permaculture farmer who five years later would marry me in Australia, was pulling up weeds in her vegetable garden above the pond. Her thirteen-year-old daughter, Kat, was at home from school on summer break, back at the solar-powered mud-brick house Barbara had built with her own hands. Out of nowhere, as she worked, Barbara was slammed by an intense, jolting, horrifying image of a bloody body. *Oh my God,* she thought, *something has happened to Kat!* She tore back up the path to the house. Intense relief. Her daughter was fine. "Yeah, Mom, I'm okay." With a reassured smile, allowing herself to relax, Barbara went back to finish her work. A momentary aberration, she decided, and put the ghastly vision out of her mind. (What would *you* do?)

An hour and a half later, a police car stopped in the pasture across the fence from the house, having negotiated the rutted two-mile easement from the highway to the top of the hill. A woman cop was standing on the pasture side of the barbed-wire fence, apparently wondering how to cross to Barbara's land. *Oh shit,* Barbara thought, *what now?* The cop shuffled, told her that her beloved cousin Lizzie, a pharmacist who lived in the nearby town of Lockhart, had been killed instantly a little earlier in the day when her car was struck at an intersection. The police, it turned out, had been delayed in bringing the tragic news by the need to go to the office of the rural electric power cooperative to get a key that would open the six gates that obstructed the easement at intervals.

As far as Barbara could estimate, her terrifying vision—an experience she'd never had before, and has not had since—must have occurred pretty close to the moment Lizzie died. That night, grief-stricken, she noted the experience in her diary, which is how I can be certain that the embroidery of memory has not played her false.

What are we to make of such testimony? Are people really connected in some mysterious way that is still unrecognized by canonical science, although familiar to almost everyone? Or are dramatic events that give rise to this conviction (and even their everyday, homely equivalents) nothing more than chance accidents in an immensely complex world? Are these just coincidences, random conjunctions that we recall only when they correspond to some vivid and shocking event? Certainly that must be true in a great many cases. It is far from obvious, though, that Barbara's appalling, singular vision can be explained away so easily.

Ten years and one month later, Barbara's farmhouse, together with much antique furniture, tools, artwork, and memorabilia stored within, burned to the ground one afternoon because a guy wire snapped. It had been holding steady an electric company transformer. The transformer exploded as it toppled into the tinder-dry Texas summer grass, spilling blazing oil and reducing to ashes the home where she'd raised her daughter. Fortunately, nobody was living there at the time, so there were no livestock, either, to be incinerated.

Barbara had no faintest premonition of this disaster.

Experiences of the former sort are often called *paranormal*, exactly because they seem to deviate from the normal, the lawful, the expectedly unexpected, which can be represented by the latter tragedy. The scientific study of apparently paranormal phenomena has been dubbed *psychical research* or *parapsychology*. The object of this often derided science—a struggling, underfunded, understaffed science barricaded on the wrong side of the gates of the orthodox scientific community—is a purported array of weird phenomena that mainstream science rejects. These include *telepathy*, or acquisition of knowledge and images from the mind of another without using customary sensory means; *clairvoyance* or *remote viewing*, where information is attained more directly, but still in the absence of any known means; *precognition*, or accurate information obtained, especially scandalously, in advance of the future event; and *psychokinesis* (PK), or anomalous actions produced by simply willing them to occur, with, once more, no known physical or energetic medium of operation. Together, these are known as *psi phenomena* (where "psi"—sounds like "sigh"—is borrowed from the twenty-third letter of the Greek alphabet because it begins the word "psyche," or mind). More recently, informational psi or *ESP* (*extrasensory perception*) is often called, blandly, *anomalous cognition* (AC), while psi-mediated action is called *anomalous perturbation* (AP). By next year the name will probably have changed again.

Why is it that just one term—psi—is now accepted for these many different oddities? An obvious question is whether these dubious phenomena do indeed have anything in common—beyond the possibility that they're all simple misunderstandings of unusual but perfectly explicable coincidences. It might especially be asked: why link PK (or AP) with ESP (or AC)? The first implies an ability to modify the brute world of matter and energy just by wishing it (an output), while the second would be an anomaly of access to information (presumably an input). This standard parapsychological decision to treat both sets of weird effects as subclasses of a single grand psi phenomenon can undoubtedly be traced back to prescientific beliefs in magical powers. Ancient magicians (and many shamans to this day) claimed powers not only to read minds and plumb the future, but also to work many kinds of physical miracles—to cure the sick, curse the wicked (sometimes lethally), affect the weather, or levitate themselves or inanimate objects. When early scientific studies were made into mesmerism, spiritualism, and other baffling claims, there was an inevitable tendency to carry forward into those studies old categories from the past.

This might not look like a very scientific basis for cool scrutiny of claims of the paranormal, but it has the virtue of not discarding observations made throughout history. Mediums—these days called "channels," or indeed, once again, plus ça change, plus c'est la même chose, "mediums"—claimed access to knowledge unavailable through customary means, but that was hardly the end of it. Their séances teemed with table tilting, manifestations of mysterious "ectoplasm" (a gauzy material that seemed to evaporate under scrutiny), strange loud clicks, and materialization of objects out of nothing. I am not for a moment affirming that these claims were valid. My point is that strange, inexplicable physical and mental powers were held to be linked—usually in the person of some magus, saint, or medium. Parapsychology retained this linkage, at least in the sense that claims of both paranormal knowledge and physical effects were studied evenhandedly, at first through observation "in the wild" (of mediumistic trances and séances, for example, and dreams or apparitions), and later in carefully structured laboratory experiments. During this history of what we might term "demystification," the alleged continuity between ESP, precognition, and PK, or anomalous cognition and perturbation, came to be understood (I believe) by an analogy with the links between ordinary perception and willed action.

Suppose we see a ball coming and hit it with a bat held in our hands. This fluent sequence seems in its very nature altogether and seamlessly integrated—unless you happen to have trouble with your eyes, as I do, or you're uncommonly clumsy. Yet the first of these activities, seeing, occurs when incoming, lens-focused light causes chemical changes in specialized cells on the retina, and these in turn send electrochemical impulses up into the brain. Our arms soon swing the bat when outward impulses activate electrochemical changes in our musculoskeletal system.

Now, these two systems—electromagnetic input and muscular-kinetic output—are in many respects strikingly different. What links them is their mutual history as evolutionary adaptations of a complex life form to the threatening world out there (via, it's true, the laws of physics). Vision is only useful if you can also grab or evade. Altering the world works better when you can see what it's like out there and then observe the results of your interventions. So, too, it is surmised, paranormal *perception* (ESP or precognition) is likely to couple to a paranormal capacity to *act* on the world (PK). This might be a false analogy (or it might be an excellent one), but it certainly seems to make good evolutionary sense—if we surmise that there are

physical laws or facts about the nature of the universe, as yet undis-
covered, that can serve as the pathway between human intentions and
the external world. Indeed, recent quantum-theoretical efforts to
account for paranormal intrusions (by former dean of engineering
Robert Jahn of Princeton University, physics Nobelist Brian
Josephson, and others) discern deep and consonant implications for
both physical and communication processes. These theories assert
that consciousness is always coupled to the physical world through
quantum mechanisms of observation that involve information rather
than energy, and is thus enabled (sometimes) to evade (just a bit) the
usual constraints of nonintentional physics.

What could this mean? Don't scientists tell us that they are on the
verge of achieving a true "theory of everything," a comprehensive
account of all the forces and interactions permitted by the laws of
physics? Wouldn't the prospect of such a total theory leave no place
for "inexplicable phenomena," such as PK or precognition? The only
honest answer is that we do not yet know. Some of the best early candidates for a theory of everything have already been bypassed in favor
of alternative string or brane hypertheories, and even those are in
question. No deep understanding of such theories is possible to the
lay onlooker. It's clear, though, that all the answers are not yet in—
and that when they do come, they might bring some major surprises.
That, after all, is the pattern of the past.

☙❧

For a whole human lifetime, starting in the mid-1930s, the emphasis of
parapsychology has been on laboratory attempts to trap these spooky
effects and draw them under the scrutiny of scientific investigation—
and ultimately, it's been hoped, to develop a powerful theory that
both explains how they are possible, given what science knows
about the world (an awful lot), and how to make use of them. The

tendency has been to ignore individual anecdotal experiences like Barbara's, since it's so hard to establish their veracity. And yet . . .

In a 2002 paper on the history of the relationship between statistics and parapsychology, a mathematical statistician, Dr. Douglas M. Stokes, observed:

> When I was teaching a course on parapsychology as a graduate student at the University of Michigan, one of my students told me that his father was knocked off the bench one day by an "invisible blow to the jaw." Five minutes later, his father received a call from a gymnasium where his wife was working out informing him that his wife had just broken her jaw on a piece of gymnastics equipment. Admittedly, this is now a third-hand case . . . [but] it would be absurd to argue with regard to such a case that men are constantly being knocked off benches by invisible blows to the jaw and that sooner or later one of these events is bound to occur simultaneously with the breaking of the jaw of a member of the man's immediate family. (16)

Frankly, I have to say I find this particular tale, by contrast with Barbara's horrific vision, unintentionally absurd (even if true), like a pratfall out of a comical movie from the 1950s or earlier. *Abbott and Costello: The Invisible Man,* say. Or Laurel and Hardy in the 1930s, slapping each other stupid. Some paranormal anomalies, to use today's terminology, found in traditional scriptures and ancient teachings strike us as no less silly. The prophet Elisha, in the biblical Second Book of Kings, was so outraged when a jeering gang of small children mocked his baldness that he summoned a pair of she-bears out of the woods and caused them to tear forty-two of the youngsters to pieces. Perhaps (as your local minister might assert) this is less a work of highly effective psychic pique than of direct divine intervention; it's hard to be sure. At any rate, most cultures have been fairly confident that sometimes people can see the future or their hidden fate in dreams or in the shapes of entrails, communicate directly mind to mind with others at a great distance, or perhaps change themselves into animals and back again. Or maybe that has always been a metaphor, a colorful way of talking about the confusions of life and our difficulties in making our way in such a baffling and often inconvenient world. As well, such images may represent a mythic transformation of what we recall from childhood, when our parents seemed literally to read our minds, when an inarticulate tantrum was enough to fetch comfort (or sometimes punishment), and when we had but to ask and food and clothing would be provided as if by magic.

The eighteenth-century Enlightenment attempted to put a stop to such superstitions and holdovers from infancy, to clear them from the minds of men and women who had long been mired and trapped by ridiculous beliefs taught to them by those who would gain from their credulity. It's ironic, then, and yet somehow perfectly reflective of human perversity, that in the middle of the nineteenth century the great booming industrial civilizations of Europe and the Britain Isles were enraptured by rumors of mediums, of heavy tables that flung themselves about in the dark under the sweating hands of the sitters, mysterious voices speaking ambiguous prophecy from floating trumpets, creepy exudations of ectoplasm emerging from mouth or vagina, gleaming in the dimness, taking on the features of the dead, impalpable hands that nonetheless groped at one's leg or seized one's hair in the dark. It was the heyday of spiritualism, a distinctly non-Christian or perhaps para-Christian form of transcendental belief emerging like ectoplasm from the very center of Victorian culture and its counterparts in the other haughty cultures of the West.

By the turn of the twentieth century, Sigmund Freud and his rebellious disciple Carl Jung were at loggerheads over the status of such weird phenomena, of the mind reading and meaningfully coincidental synchronicities that seemed to attend the couch, and the new science of psychoanalysis. Solemn Societies for Psychical Research were founded in Britain (1882), America (1885), France (the Institut Metapsychique International, 1919), eventually Austria (1927), and elsewhere, their early officers and presidents often men and women of significant scientific standing, Nobelists even. The dead spoke to them, usually in muffled riddles or platitudes, yet somehow reaching poignantly into their hearts. Perhaps the demystifications wrought by science had parched their spirits, as many opponents of science still maintain it must. In any event, the table rapping and the mind reading lost their charm, eventually, attended by gales of laughter as Sir Arthur Conan Doyle, creator of Sherlock Holmes, was taken in by fake cut-and-paste pictures of fairies and other comical sideshows.

It was not until the 1930s that the brash new face of experimental, somewhat behaviorist psychology was turned upon the alleged mysteries of the paranormal. Not that psychical research had halted; its upper-middle-class adherents continued their dutiful searching for cross-correspondences in the utterances of spirit mediums, hoping to prove that the dead spoke in ciphers that might be decoded ingeniously if only you could find all the bits and pieces of their messages scattered between the mediums. The mood of the 1930s, after the brutal, pointless horrors of the First World War and the ruinous dislocations of the Depression, was cheered by the news that a plant physiologist turned psychologist turned *para*psychologist at Duke University in Durham, North Carolina, Dr. Joseph Banks Rhine, had trapped psychical phenomena in his laboratory and given them a name: ESP, or extrasensory perception. Rhine's germinal book of that title had been released in 1934—*Time* magazine snidely noted that it "became the brief rage of women's clubs all over the U.S."— and the news from Duke was getting through to the public. Within a few years, this tall, stocky, impressive former marine had added psychokinesis to his tally of attested laboratory wonders. In fact, he had worked on this mind-over-matter phenomenon almost from the outset, but that particular astonishment he quite sensibly held in reserve until after ESP had proved itself acceptable, indeed enticing, to a shell-shocked and disenchanted world. By a quite curious irony, Dr. Rhine was color-blind (along with his hammertoes, this shortcoming had almost barred him from joining the U.S. Marine Corps in 1917, until he bulled his way in). Worse, an attack of measles when he was in his late teens had destroyed his sense of taste and smell, partially deafened him in one ear, and damaged the sight in one eye (Brian, 13, 15). One must wonder, were these deficits a spur to his developing obsession with the *extra*sensory? (For what it's worth, I've been half blind myself since childhood, lacking the usual three-dimensional gift of depth vision; it has to give you pause.)

Perhaps the key distinction between Rhine's school and his spiritualist and psychical research predecessors is one of method—or, even more crucially, cast of mind. The psychic researchers at the end of the nineteenth century resembled historians or geographical explorers. They accumulated anecdotes, often in impressive detail; they attended and observed séances and other manifestations of the purported paranormal, taking copious notes and reducing their observations to narrative. Call them *storytellers,* and their laboratory successors *bean counters.* This is glib, admittedly, but I think it captures a major difference in the two kinds of approaches. What the one retained in human warmth and meaningfulness—not to mention thrilling creepiness and hints of transcendence—the other made up for (or hoped to) in rigor and repeatability, the hallmarks of science. The storytellers became the old way; the bean counters were the new. For a while. For half a century, in fact. In the additional third of a century since the mid-1970s, research into the paranormal has gradually, by fits and starts, accommodated elements of both approaches.

Curiously, Rhine the scientist began his studies as a would-be divine. The religious instruction at his first college did not appeal to him, and he followed his schoolteacher neighbor, Louisa Weckesser, to the College of Wooster, where he found ministerial training equally unpalatable. At twenty-two, at the end of 1917, he joined the marines. In 1920 he and Louisa married and they picked up their graduate studies at the University of Chicago, a campus that had been galvanized by the arrival of behaviorist John Watson, an extreme reductionist who regarded mind and soul as remnants of ancient mythology, to be ruthlessly stripped out of the human sciences. Rhine was impressed, but not altogether persuaded. In 1922 he attended a meeting in Chicago where physician and celebrated author Sir Arthur Conan Doyle forcefully presented his testimony concerning the truth of spiritualism. Rhine was unconvinced by Conan Doyle, but he did take the trouble to read a book recommended at the lecture, *The Survival of Man* (1909), by notable British physicist Sir Oliver Lodge. This led him to the bulletins of the Society for Psychical Research and their reports of recent work in telepathy and clairvoyance.

Rhine's evenhanded biographer, Denis Brian, insists that "although Conan Doyle and Lodge were spiritualist, Rhine never became one" (23). His growing interest in what later would be called nonreductionist psychology led him in turn to the head of Harvard's psychology department, William McDougall, Watson's most energetic critic and opponent. McDougall was president of the Society for Psychical Research in 1920. "It was with a definite purpose of undertaking investigation into possible unknown capacities of the mind, the so-called psychic powers, that I first got in touch with Professor McDougall," Rhine told Brian (26). And as a result, after stints as a plant physiology instructor and lecturer in New York and then West Virginia University, he investigated the celebrated Boston spiritualist Mina "Margery" Crandon and her husband, Dr. Le Roi Goddard Crandon, favorites of the American Society for Psychical Research, found them spurious, and denounced them, to Conan Doyle's fury.

In his 1927 report on a fake Margery séance, in the *Journal of Abnormal and Social Psychology*, Rhine wrote that his motive had not been to attack the spiritualists, "but to help distinguish sharply between such fraudulent activity and the genuine basis of psychic research which continues to hold our respect and engaged our interest."

Rhine and his wife had moved to Boston to work with McDougall, who, unfortunately, immediately left on a world trip with his family. Academic life for the underlings in those days rather resembled the way it is today: in 1927 Rhine was obliged to work through the summer selling aluminum utensils from door-to-door, learning the skills of salesmanship that would serve him well in subsequent decades as he sold his new science to the world. Finally, when McDougall moved to Duke University to head the psychology department there, the Rhines followed and began their psychic career in earnest rather inauspiciously with an investigation of a supposed psychic horse, Lady Wonder, who could tap out simple arithmetic and "read minds." It took a surprisingly long time to establish that what Lady was reading was conscious or unconscious signals from her trainer and onlookers. (Actually, though it's easy to laugh at these blunders, if psi is a reality, there is no obvious reason why other species should not share that prowess. DMILS, direct mental interaction with living systems, is currently a thriving subdiscipline of parapsychology.) In any event, Rhine retained an interest in animal psi and with his associates investigated the possibility that pigeon homing was in part a psychic phenomenon. The question of postmortem survival continued to exercise him. It was not until 1930 that Rhine borrowed from the French Nobel Prize–winning physiologist Charles Richet the idea of using playing cards in opaque envelopes as a test instrument for probing psychic powers. The shift had been made from an essentially anecdotal, narrative, storytelling method to the bean counter's stricter and more parched annotations of data.

Rhine abandoned ordinary decks of playing cards, switching to cards numbered zero through nine. By chance, you should be able to guess close to one in ten of these correctly. Make enough calls, and the random variations due to chance will fall away into the background, making any correct excess guesses stand out. The problem with using numbered cards is that people tend to have "lucky" and "unlucky" numbers, or favorite numbers associated with birthdays, conscious and unconscious detritus that clings to the undersurface of the mind as one is trying to make clairvoyant or telepathic identifications of the hidden cards. So Rhine asked a colleague, perception expert Karl Zener, to design five simple forms that might be printed on cards and used as the basic tool for psi testing. Zener provided the templates for what would become the most famous symbols of ESP—square, circle, cross, star, waves. (Zener later fell out with Rhine, and rather resented the way his name had become attached to psychic phenomena.)

Within a few years, Rhine had gathered about him at Duke a small, dedicated team of investigators and some remarkably gifted experimental subjects among the hundreds of students they had tested. His most remarkable early gifted subject was Adam J. Linzmayer, an economics undergraduate who believed that he and several other anonymous students had been recruited by an equally anonymous visiting professor to form a covert think tank aimed at solving the problems of the Great Depression. Unfortunately, the professor vanished after several secret meetings, and that was the end of it. For all I know, this story might be true. It illustrates beautifully the extremely vexing interpenetration, within the history of psychic research, of apparently preposterous claims and equally astonishing test results. If you'd been told twenty years ago that the United States Central Intelligence Agency and various arms of the U.S. military were conducting secret operational research into psychic powers, funded by Congress, would you have believed these tales? But as we will see shortly, that was exactly the simple truth. In any event, Linzmayer startled Rhine with bursts of remarkable accuracy under conditions both controlled and, let us say, less controlled.

Critics have pounced on the here-it-is, there-it-goes nature of such claims. In May 1931 Linzmayer correctly guessed nine Zener cards in a row, something likely to happen by chance about one time in 2 million. The next day, he did it again. Rhine ran him through 300 calls, and Linzmayer got 119 correct. By chance, given that each call has one chance in five of being right, he should have had only 60 plus or minus perhaps 14. In statistical jargon, his score was 8.5 standard deviations from the mean. This is spectacular.

But here is the problem with using serial card guesses as a method for probing psi—it's as boring as hell. Rhine knew that from the outset, even though he persisted with that methodology because of its merits in providing ease of statistical analysis. Nonetheless, Rhine understood that he could not press Linzmayer indefinitely to keep calling one dull card symbol after another. To break the pattern, on June 3 he drove Linzmayer into a forest, stopped, and ran through a series of calls. Linzmayer got 15 correct in a row. They paused for a time. In the next 6 calls, he was wrong 4 times. To obtain so many correct calls out of 25 is staggeringly improbable. The question is, how likely is it that Linzmayer, sitting in the passenger seat of an old Essex, could find clues—reflections, unconscious indications from Rhine himself—by normal but subtle means? Skeptics will have no doubt of the answer. Rhine was exultant. In 1931 he tested Linzmayer again, after a falloff in scoring, and his remarkably gifted subject managed an average of 6.5 out of 25 in 945 calls. This is still impressive evidence for ESP. Meanwhile, 24 other subjects made 800 calls, also with an average of 6.5 correct. The following year, Linzmayer repeated his moderately good scores, rising to 6.8 out of 25. But by the spring of 1933, in an exhausting series of runs producing 2,000 trials, his score dropped to 5.9, and kept falling.

Meanwhile, eighty students scored at chance level. Worse still, a factor called "psi-missing" started to show up. In a test of the Zener-card sort, it is most unlikely that respondents will guess the wrong answers much more than four-fifths of the time. Paradoxically, you can be *wrong* significantly more often than this only if you *know the right answer* (paranormally, and probably without conscious awareness), but deflect your response (for whatever unconscious motive). Yet why would anyone purposely psi-miss? Plenty of explanations have been proposed over the years. One of the most commonly accepted is the psychoanalytic notion that we desperately fear intrusions on our protected inner life, and that we dread drastic change. It is quite feasible that the exercise of paranormal cognition, especially when its consequences could be life changing, jolt the rational mind into a deep-set mechanism of denial and defense. So psi-missing might be a kind of complicated skill that humans learn, in our culture at least, to save us from too much, too quickly, and by too weird and threatening a channel.

Alternatively, it might be a mental quirk like making a slip of the tongue. On the face of it, it's just as astonishing that anyone would utter the wrong word rather than the one they meant. Again, many explanations have been advanced for this common experience. Psychoanalysts find evidence here for the momentary breakdown of unconscious repression, so that a disturbing word slips out when another that resembles it is "on the tip of your tongue." Cognitive scientists propose a less suspicious theory of the mind's inner workings: lapses occur in such circumstances rather as they can during typing when fingers strike the wrong key, turning "vat" into "cat." Psi-missing, which was discovered empirically, could be due to either of these causes, or both, or indeed to some other motives not yet isolated. None of this means that we cannot readily detect even so surprising a phenomenon as psi-missing once it has been observed in the lab. We can even look for its correlative psychological states (some of which seem to include anxiety and introversion).

A new superstar emerged in 1932, under the auspices of Rhine's student and later colleague Joseph Gaither Pratt, another former

student for the Christian ministry who had switched to psychology. Pratt worked with a shy student, yet another divine, named
Hubert Pearce, who was convinced that he and his mother both had
psychic powers; Mrs. Pearce had once levitated a heavy oak table
that floated upward from the floor even though several men pressed
heavily upon it until it snapped across the middle. This is the kind
of storytelling background that precedes many of the bean-counter
programs and their dutifully scientific subjects. You see it today in
remote viewing projects replete with entropy gradient studies, fuzzy
set evaluations, statistical measures up the wazzoo, and viewers who
when pressed reveal their frightening experiences out of the body
during near-death episodes, and the like. It's worth keeping that in
mind. If psi exists, if it's an evolved function of the human
genome—indeed perhaps of other species as well—no more nor
less remarkable than our capacity to see distant objects with our
eyes, then we can guess in advance that it is not specialized or opti-
mized for guessing cards or pictures in a laboratory.

ೲ

There are three key questions to be asked about psi phenomena:
First, *do such phenomena really exist?* Or are telepathy, remote
viewing, precognition, psychokinesis, and the other apparent paranormal anomalies nothing more than coincidence, misunderstood
normal processes of intuition, mythology, or fraud? Second, if they
do exist, *what function do such phenomena subserve?* Or to put it in a
slightly more familiar but risky way, what is their purpose? (The
reason that question is risky is precisely because "purpose" is normally a word we reserve for conscious human intentions, and evolution doesn't have any. On the other hand, the capacity to devise and
act upon a purpose is itself evolved, and hardly restricted to
humans.) And finally, assuming that such phenomena exist, even if

they don't have a purpose, even if they are just accidental by-prod-
ucts of some other natural mystery (like the accidental flashes of
phosphene light seen by astronauts when random cosmic particles
slash through their unprotected retinas), *what causes them?* How do
they come about? What is the physics of psi?

We'll consider these questions from many angles, searching for
some (at least tentative) answers. To start with, let's proceed with that
founding question: Is there anything to it? Some persistent seekers
after the paranormal, in a return to the spirit of the spiritualists, report
much stranger events than anything studied by Rhine, including levitation, dematerialization of objects, and all kinds of exotic phenomena
that are rarely studied during the workweek by establishment parapsychologists. All that an unbiased observer can do is listen politely,
frown, and put the story into a folder marked "Paranormal Anecdotes," alongside tales of people who see auras glowing softly around
human bodies, and reports of the Virgin Mary walking and bowing
on the roof of a church in Zeiton, Egypt, witnessed and apparently
filmed by thousands, not to mention scads of other anomalies and
apparent absurdities well outside the experience of almost all of us.

At the end of his long, detailed study of the vexed methodological
issues involved in studying the purported paranormal, the statistician
Douglas Stokes comments that his "own tendency to believe in psi is
based more on acquaintance with particular spontaneous cases than
on the statistical evidence that has been assembled. In all probability,
however," he adds ruefully, "it will take evidence of the latter type
rather than the former to convince a skeptical scientific community of
the existence of psi" (56).

೧೨

The fact that people have believed in what we now call paranormal
phenomena for tens of thousands of years is not sufficient answer in

itself. It's very hard to wrap our minds around the stories ancient people told one another about the world, its origins and nature, the obligations that knowledge imposes upon us. Thousands of years ago, most people believed things that are so weird it's hard to grasp how they managed it without laughing. Then again, many of us continue to have ardent faith in a small set of those claims, while declining to accept most other traditional opinions or dogmas. It's likely that more people today have a working knowledge of astrology than they do of scientific astronomy. Human societies have sustained for tens or hundreds of thousands of years all kinds of beguiling but incorrect claims about the nature of the world we inhabit— notoriously, that the world is flat, at the center of the universe, and that the sun, moon, and stars rotate around it; that a gusty wind in the leaves is proof of the spirit dwelling in the tree; that life was hand molded, species by species, by an Intelligence rather like humans but seriously more awesome; that illness is caused by sorcerous malignity; and that dreams carry us into worlds as real as this, where the dead continue to walk and speak and intervene in our lives.

It is a vast and shadowy heritage, embedded in the very language that we use—"the sun also rises"—and yet we now know that it is almost entirely false. Well, except for the 43 percent of Americans who still doughtily insist that "Darwin's Theory of Evolution" is blasphemous balderdash.[1] Ah, yes, "Darwin's Theory of Evolution." It has a wonderfully antique nineteenth-century flavor, complete with capital letters on every noun. Does any newspaper routinely refer, in an article on cardiac treatment, to "Harvey's Theory of the Circulation of the Blood"? The irritating aspect is not just that evolutionary analysis is now the basis of all biological studies, but that in the real world Darwin's brilliant but limited role has long been subordinated by our increasingly exact understanding of genetics. Oddly, no newspaper speaks, in an article on DNA, of "Mendel's, or Morgan's, Theory of Genes." This disparagement of an immensely exfoliated set of related sciences by reducing it to One Dead White Male's Opinion reaches its absurd nadir in headlines such as "God vs. Darwin." Deft and catchy, but I would hope to see, at least once, a headline (it's rather too long, I know) proposing "Ancient Ignorant Guess vs. Darwin, Mendel, Crick and Watson, and the Human Genome Project."

It might not be possible for those of a scientific cast of mind to appreciate these deeply divisive antique explanations for the world, to reach back into history and prehistory, to reach outward ecumenically to cultures whose beliefs seem frankly crazy or offensive. Some regard that ambition itself as nothing better than a wishy-washy politically correct denial that certain sources of knowledge actually are more equal than the rest. At any rate, against that dire background of persistent and heartfelt confusion, one is obliged to ask: Is belief in psi nothing better, after all, than another living fossil from those stygian, blundering blind guesses that ruled the world long before the Dark Ages?

After a brief easing of traditional skepticism in the 1970s, which saw journals like *Nature* and *New Scientist* publish papers claiming to substantiate the paranormal, the gates of mainstream science have almost closed tight again. Nearly thirty years ago, in a spectacularly apt declaration of the renewed freeze on psychic investigation, Dr. John Archibald Wheeler of the University of Texas demanded in 1979 that the prestigious American Association for the Advancement of Science (AAAS) shun parapsychology as mere pseudoscience, barring its practitioners from the "workshop of science." He was like an angry Old Testament deity, casting out the sinners and placing angels with a flaming sword to barricade the gates of Eden—or in this case, at the gates of science itself. Wheeler was so outraged by the measure of sympathy that parapsychology had been attracting, limited though it was, that he called for the AAAS to withdraw affiliate membership from the sober academic Parapsychological Association. In March 2006 the shockwave of this outburst was still ringing in some ears. *New York*

Times science writer Dennis Overbye claimed, in an essay titled "Far Out, Man, But Is It Quantum Physics?" that this expulsion had actually been carried out: "The parapsychologists were booted from the American Association for the Advancement of Science 30 years ago." The newspaper later corrected that false claim, but made no apology.

Bear in mind that Wheeler was the very contrary of a hidebound conservative. In the 1930s he helped Niels Bohr perfect quantum theory. Thirty years later he was the one who adopted the catchy term "black hole" for a collapsed region of spacetime from which nothing, not even light, could escape. He was a leading contributor to the study of gravitation and high-energy physics. So for half a century his brilliant, innovative intellect was at the forefront of the physics revolution; his graduate student, the late Richard Feynman, won the Nobel Prize for inventing key parts of quantum electrodynamics, a theory that has produced the most accurate prediction of a physical quantity in the history of science. Wheeler's 1979 assertion was that parapsychology—still no better than a fledgling science of the anomalous, despite a century's concerted effort—had yielded not one single "battle-tested" discovery that could be replicated by skeptical or inquisitive scientists. Until it could do so, he argued, it was little more than a sham, and did not deserve to stand shoulder to shoulder with authentic science—let alone exploit the grudging tolerance of authoritative bodies such as the AAAS in order to garner funds for its dubious research, at the expense of orthodox scientific projects.

Yet there undoubtedly *had* been scientists of high caliber studying the paranormal (or, better, the anomalous) during the decades preceding Wheeler's outburst—and indeed in the decades that followed it. These were specialists who were initially from outside the self-selected ranks of parapsychologists. Were they shaken by this denunciation? The most striking evidence gathered by non-parapsychologists during the last quarter century was conducted at Princeton University. Until recently, Dr. Robert Jahn (pronounced somewhere between "jarn" and "john," not "yarn") was Princeton's emeritus dean of engineering. One of the world's leading experts in advanced aerospace propulsion technology, he worked with major industry and government research laboratories in fluid mechanics, ionized gas physics, and plasma dynamics. From 1979 until its closure at the end of 2006, Jahn also directed one of the most remarkable and impressive bodies of public research into psychic phenomena. Jahn's PEAR team—Princeton Engineering Anomalies Research—came as close to proving the reality of psi as you could imagine (although a recent large-scale international attempt to replicate one of its key findings has failed). Jahn and his principal associate, developmental psychologist Brenda Dunne, showed astute political horse sense in avoiding what they term "gee whiz" experiments, flashy psychic bombshells of the Uri Geller spoon-bending variety. Even so, in a 2005 retrospective, anticipating the closure of their program, Jahn and Dunne recalled wryly:

[T]he prospect of mounting a research program of a scale and character competent to render definitive answers to the host of strategic and philosophical questions swirling around such an investigation was daunted by a recalcitrant university administration and a dearth of scholarly colleagues willing and competent to collaborate in such an enterprise. More serendipitously, however, there was also at that time a compensatory eruption of intellectual, emotional, and not least, financial encouragement from a number of powerful supporters outside of the university who were unflinching in bringing their stature, influence, and fiscal resources into play. . . . After a tedious period of frequently frustrating and occasionally amusing negotiations within the university, the program was authorized and officially launched in June 1979. Minimal laboratory space was carved out of a storage area in the basement of the Engineering School complex, which to this day remains the storied technical and social home of the fully international PEAR operations.[2]

In an extensive series of stolid, conservative tests using mainly three distinct classes of data gathering, they looked for very small but highly consistent deviations imposed by deliberate intention upon a background of random "noise." Because psi effects in laboratory studies are almost always found to be at the very fringes of detection, it seemed necessary to concentrate and distill their transitory presence (if any) into an accumulating profile.

The principal claims that parapsychology has made in a century and a quarter of psychical research are that psi phenomena are real but sporadic, low in efficiency, and resistant to normal reinforcement techniques of training. Although some researchers claim to have uncovered certain "psi-conducive states," their results nevertheless always remain at the margins of visibility. It was on this frustrating foundation, using applied information theory, that the Princeton group (and most other reputable parapsychologists) built their provocative databases. The three general protocols explored by Jahn press on a single disturbing claim: that the intentions of a human consciousness can *directly* influence the world or another consciousness, or be influenced by it, without any medium or mechanism of mediation known to science. Two of the experimental regimes tested for psychokinesis. The third was even more distressing; it sought to acquire information from locations that were remote in time as well as space. Often, this third protocol attempts to perceive and record *future* circumstances that are not predictable, even in principle, at the time of the experiment.

In the first of the two PK arrangements, sophisticated machines were used to generate and record streams of binary numbers—basically, strings of pluses and minuses—from a commercially available "white noise" source. Such a machine draws samples from the electronic hiss of a diode at regular intervals, usually much faster than the human mind is capable of noticing or of responding: anywhere between ten and ten thousand times a second. If the random sample is greater than average on one pass, a positive unit of data (a *bit*) is recorded. If less than average, a negative bit is recorded. On the subsequent pass, this mechanism is reversed, thus providing a string of pluses and minuses that is immunized against any positive or negative bias at the diode level.

With this higher-level string, samples of between twenty and two thousand bits are taken, and the value found is recorded and sent out to the visual display terminal. (Usually the count is close to half the number of bits in the sample, since the inherent probability involved is one in two. That is, you would expect to find around ten if the sample checked through twenty bits, or one thousand if the number of bits examined was two thousand—just as you would if you were tossing coins, around ten heads or tails from every twenty coin flips.)

Each such sample was termed a *trial*. The basic task of the operator, sitting relaxed but alert in front of the machine, was to shift between three distinct states of mind, according to a careful, preset tripolar schedule. In one condition, an attempt is made to nudge the random noise outcome higher than the mean value expected by chance. In simple terms, this involves "persuading" the device to produce more plus trials than minus ones when that is the desired outcome.

But there's always a theoretical risk with these machines that some unknown but natural factor might skew the output high or low. So to balance that, if it were to occur, a second state of mind is invoked: the wish that the machine will now produce more *minus* votes than plus ones. Of course, the alternation from one desired outcome to the other is always declared and recorded in advance, and the machine is constructed in such a way that tampering by the operator (or, indeed, the experimenter) is virtually impossible.

So now we have two sets of trials, one aiming high, the other low. It is a mark of the ingenuity and doggedness of Jahn and Dunne that they went on to ask for a third interleaved set of trials. Now the aim was to get the machine to run out its string of bits in a perfectly *normal*, chance manner. You can take this to mean one of two things: either the operator is doing *nothing* during these specific trials, allowing the machine to function free of any alleged psychic input, or the operator is trying to bring about a condition that is even more "normal" than normal—a kind of artificially imposed, pure bell-shaped curve. Either way, this baseline condition serves as an elegant control for the other two outcomes.

These random event generator (REG) experiments were made in *runs* of fifty trials at a time (although some runs have used one-thousand-trial groupings). The runs, in turn, were aggregated into blocks of fifty or one hundred, which thus comprise twenty-five hundred or five thousand trials, or, in the case of one-thousand-trial runs, into three-thousand-trial blocks. These groupings are called *series*, and are neatly accessible for statistical treatment.

Other secondary factors were explored during these decades of research. The rate of bit sampling can be varied. After declaring the intention to do so, operators may decide on a whim whether to aim either high or low (just so long as at least five consecutive runs are made in each condition, otherwise certain lethal statistical errors can creep in), or can follow a random machine-generated set of target requirements. Adding these options made the experiments more user-friendly without sacrificing rigor, and allowed PEAR to look for useful hints in the best way to optimize paranormal effects.

A twist on the diode-driven REG technique is to switch in an alternative source of randomness, in this case "pseudorandom" numbers created directly by a computer program. Strictly speaking, these streams of data are entirely determined by the equations that churn out the numbers, so they are still less open to human mental control.

Extraordinarily, some parapsychologists who have tried to modify pseudo-number output claim that these bit streams, too, differ from the results expected by pure chance[3] (although PEAR found only chance results with pseudorandom numbers, after early errors were corrected).[4] If it is difficult to grasp the idea that mental intentions can enter a diode and influence its real-time production of random hiss, it is almost impossible to get your mind around the thought that PK might distort the output of a computer algorithm. The most likely place for such an effect to intervene is in the choice of the "seed number" that has to be fed into the computer to start the pseudo-random sequence running. Dean Radin has commented to me: "The claim is not that PRNG [pseudorandom number generator] outputs are altered from what they would otherwise produce, but rather that the seed numbers are selected to provide PRNG outputs that conform to the intended direction. The former would require that a computer operate in an anomalous matter, and to my knowledge there's no evidence that this has ever happened." Still, this unexpected and apparently absurd result is one motive for another school of anomalies researchers to try an entirely different theory of psi, one that involves precognition rather than psychic force: DAT, or decision augmentation theory, to which we'll return.

And the implications, if this thought is entertained, are truly outlandish. For in some sense it requires the operator to know in some detail (unconsciously) the stream of ones and zeros, pluses and minuses, that a single seed number will produce from a complicated algorithm. The task is rather like guessing whether the next million digits of pi contain more odd or even numbers! One begins to wonder if success in such a task therefore implies that the operator must start by tapping into some collective mental supercomputer, a fabulously paranormal conjecture even without reckoning on PK. It is easy to see why skeptics would rather believe in any other explanation, including massive fraud, than place their trust in these appalling results.

A second PK protocol was more intuitively, viscerally appealing. Jahn had a huge array of pegs, row upon row of them, mounted behind glass in a wall display three meters high and two meters wide, forming what is dubbed a *random mechanical cascade* (RMC). The lower half of this huge vertical pinball game contains nineteen transparent tubular bins. A container at the top is filled with nine thousand polystyrene balls half the diameter of Ping-Pong balls. When the RMC is activated, these balls pour out of a funnel at the top for about twelve minutes, falling and tumbling, striking the pegs as they dribble downward toward the collecting bins. Each ball's path is a "random walk" through the descending array of pegs.

It has been known for a century that when thousands of balls work their way down through a suitably designed RMC and pile atop one another at the bottom, they produce a close approximation to the familiar bell-shaped curve known as the "normal distribution" of random binary events. Most of the balls, after all, will tend to fall more or less straight down. If at first they stray to one side, they are just as likely to trend back again into the center. But a few will be kicked randomly somewhat farther to the right (or left) by their chance encounters with the pegs, and of those a still smaller percentage will shift to the extreme ranges of the display.

A PK operator's task is clear-cut: he or she must try to skew the torrent to the right, or to the left—or, in control mode, into the regular chance bell curve. More than twenty amateur psychics, none of them "psi superstars," finished a total of 87 three-way series, for a total of 1,131 tripolar runs. Their results, combined, are statistically significant with a chance probability of less than one in ten thousand—as are 15 percent of the individual series and nearly two-thirds in the intended direction—but showed an intriguing tendency to score at a significant level only when they were trying to skew the curve to the left. Jahn and Dunne explain that the effects as displayed in the diagram below "are derived from the differences in bin populations among the three intentions in given sets of runs, rather than comparisons with theoretical expectations, which are not readily calculable" (Jahn and Dunne 2005). It's a way of displaying the tripolar difference that hides the skew to the left, but arguably does better in showing the underlying pattern of distinctiveness in *intention*.

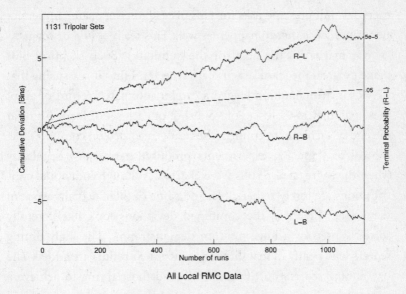

Figure 1-1. Random Mechanical Cascade. Total deviations, displayed by PEAR in a format maximizing differences between High, Low, and Baseline cumulative scores.

Is there some natural explanation? Perhaps the pegs, or the balls, are worn more on one side than the other? The tripolar design counters such an objection, for any gain in one direction due to a subtle machine bias will be offset when the target condition is reversed. (This seems to be an elementary point that eluded skeptical critic Dr. Robert Park, who was cited in *Nature* on the day the PEAR laboratory shut down as explaining that "if you run any test often enough, it's easy to get the 'tiny statistical edges' the PEAR team seems to have picked up. If a coin is flipped enough times, for example, even a slight imperfection can produce more than 50% heads" (http://www. nature.com/news/2007/070226/pf/446010a_pf.html). Park failed to notice that when he throws his slightly biased coin, hoping for an excess of heads, not only must the slight imperfection make it happen that way in more than 50 percent of throws, but when he switches every few minutes and asks for tails instead, the very same imperfection must cause that to happen as well. This really *would* be magic.

Or maybe it's the heat (or the humidity)? Such skeptical bids make no great mechanical sense, it's true, but I find it reassuring that PEAR's instrumentation logged a careful automated record of all of these parameters—plus a Polaroid shot of each set of loaded bins at the end of each run, to confirm the results recorded on computer file.

All of these PK experiments produced a revealing regularity when the aggregated results were charted. Although individual contributions are rarely remarkable, and some of them actually work in the wrong direction, the combined deviations look like carefully sorted subsets of a huge random demonstration. The high-aiming scorers tend neatly in that direction when their results are added. The low-aiming scorers do the converse. And those trying to achieve a pure-chance baseline yield an even more bland result than one would expect by chance! Only when all three sets of data are combined do the scores follow curves predicted by probability theory. "This process," Jahn and Dunne note, "bears some similarity to the 'Maxwell Demon' paradox of the kinetic theory of gases that is precluded by the second law of thermodynamics. In classical physical theory, no completely isolated system can spontaneously reduce its own randomness or 'entropy.' Thus any such preferential sorting of scores would itself constitute an anomaly."

What of having a number of operators simultaneously "will" the balls to fall in a given direction? Regrettably, Jahn never tried two thousand people in action at the one time. He did, however, try what he terms "superposition" of a few operators at once, as well as "separation," where the operator tries to induce the effect while at some distance from the lab. Jahn and Dunne are slightly evasive on this score: "While the accumulated data are much less extensive . . . it appears that anomalous results can appear, but in patterns that are not simply linear combinations of the separate signatures." A "signature," as mentioned above, is nothing more mysterious than a graph showing an individual's accumulating experimental record. These, to Jahn's eye, prove to retain a certain persistent identity over time and despite variations in the experimental design. Critics fail to see these patterns.

A separate program of experiments "has been established in which given operators address the REG and RMC tasks from various remote locations, ranging from a few miles to global distances. As with the co-operator experiments, various anomalous signatures continue to be observed, some of which are consistent with the local efforts, and some of which show characteristic differences." In short: *something* outrageous was happening.

Most extraordinary of all these protocols was what Jahn termed *precognitive remote perception.* It is a version of a method that gained considerable notoriety in the 1970s when it was promoted as "remote viewing" in the book *Mind Reach* by Russell Targ and Dr. Harold Puthoff, two Stanford Research Institute laser experts who championed Uri Geller (to the detriment of their reputations) and helped create the classified military/CIA psi program that was revealed publicly in 1995.

The nub of the method was akin to controlled fantasy. A subject went into a shielded room while a "beacon" agent traveled to one of a number of possible target destinations selected at random. After a while, the would-be percipient began to imagine what the distant location was, listing and sketching as many details and features as possible. Later, a panel of judges would evaluate the fit between remote description and each of the possible locations, and then rank-score that target set of random alternatives drawn from the target pool. Remote viewing enthusiasts claimed that matches were much more numerous than could be ascribed to chance coincidence, and indeed often revealed an eerie accuracy that eluded narrow statistical formalization.

Within a few years, these reported results had been savaged by two skeptical psychologists, David Marks and Richard Kamman. Clues inadvertently left in the transcripts had permitted judges a higher-than-chance opportunity to identify the correct locations. Targ and Puthoff immediately scoured out all the clues they could find, resubmitted their transcripts to new judging panels—and still got significant results. The critics took another look and found to their own satisfaction that enough clues and cues still remained to explain the non-chance results. Neither side conceded that the other had won the debate.

Brenda Dunne was involved in remote perception experiments before Jahn set up his Anomalies Research team, so she brought with her a body of informed knowledge on techniques ironed out in the 1970s. As a result, PEAR's remote viewing protocol attempted to guard against familiar, well-grounded criticism by making all of its remote perception data amenable to explicit computerized statistical assessment. As well as providing the now-traditional impressionistic report, the percipient filled out a checklist of thirty binary descriptors. Questions included the following: *Is the central focus of the scene predominantly natural, that is, not man-made? Is the scene predominantly dark—for example, poorly lighted indoors or nighttime outside? Are steps or stairs prominent (excluding curbs)?*

Targets were chosen in two ways. In the first, a randomizer selected from a large pool of prepared locations. Alternatively, an agent was permitted to visit some distant place (typically thousands of miles away, perhaps in another country) and choose from the unpredictable options available there. This last approach undoubtedly raises some analytical difficulties, which were seized upon quite properly by critics. Since those involved in PEAR's perception experiments usually know one another, they can estimate (if only unconsciously) what features of the landscape are likely to attract a friend's notice. Dunne disputes the weight of this objection; her statistical analysis shows no difference in the levels of significance obtained under the two alternative conditions. Still, it must be granted as the most obvious point of potential weakness in all the PEAR experiments.

Estimating the degree of success or failure is not merely made easier by forcing results into computerized binary form. It is also made more precise and protected from certain bias hazards. For instance, a computer search of the prepared target pool shows that more than half the available locations are outdoors. So the probability of getting a positive result by guessing at an indoor setting is less than one in two—to be precise, it is 39.7 percent. There are trees in 64.5 percent of locations, glass or windows in more than 70 percent, but only 15.4 percent of the scenes are judged "oppressively confined." A quick computer evaluation can scale or "normalize" each thirty-point report to take this set of biases into account and assign a clear probability rating to its degree of match or mismatch.

What's more, an empirical control can be generated in the same way for the whole aggregated series of remote events. Using 334 scores, PEAR's computer deliberately mismatched calls and targets, deriving 42,000 control scores. Graphed, these "false" results showed a very impressive correlation with chance expectation. By contrast, the number of correct bits of information garnered by PEAR's operators exceeded the normalized chance result by some 15 percent. This is spectacular in terms of laboratory psychic success (which, when it shows up at all, tends to be between 1 and 0.02 of a percent above or below chance), and Jahn estimates the probability of its occurring by accident at about one in 50 billion.

In their 1988 book, Jahn and Dunne included vivid examples of remote perception reports, together with photographs taken by the remote agent at the scene. Conveying the variety of possible targets, these include a railway station at Glencoe, Illinois; a high terrace in Milan; the ruins of Urquardt Castle, Loch Ness; a startling bridge on the Danube, in Bratislava, Czechoslovakia (described a day in advance, at a distance of fifty-six hundred miles, with no less startling accuracy); and the following perception, quoted in full, where the random target destination—again, not visited until the *following* day—was the huge radio telescope at Kitt Peak, Arizona: "Rather strange yet persistent image of [agent] inside a large bowl—a hemispheric indentation in the ground of some smooth man-made material like concrete or cement. No color. Possible covered with a glass dome. Unusual sense of inside/outside simultaneity. That's all. It's a large bowl. (If it was full of soup [the agent] would be the size of a large dumpling!)" At the risk of harping on the point, bear in mind that these are imagined descriptions *of events that have not yet occurred, places not yet visited.*

Can this stuff be taken seriously? If it is not a fairly pointless and silly fraud, might it be a colossal fluke after all? Anyway, can statistics tell us anything about the real world? Haven't we been warned often enough about "lies, damned lies, and statistics"? Certainly, the accumulated research findings still have not converted the die-hard skeptics. One critic who has himself tried some psi experiments with mixed results is Stanley Jeffers, of the Department of Physics and Astronomy at York University, Toronto, Ontario. In 2006, in a retort to Jahn and Dunne, he tried to undermine their control procedures.

One characteristic of the methodology employed in experiments in which the author has been involved is that for *every* experiment conducted in which a human has consciously tried to bias the outcome, another experiment has been conducted immediately following the first when the human participant is instructed to ignore the apparatus. Our criterion for significance is thus derived by comparing the two sets of experiments. This is not the methodology of the PEAR group[,]which chooses to only occasionally run a calibration test of the degree of randomness of their apparatus.[5]

York Dobyns of PEAR has disputed this critique in a flurry of statistics.[6] The discussion, as so often with psi, continues without apparent clear resolution.

Curiously, those top-ranking scientists who bothered to look at the Princeton figures are by no means unanimous in either applauding or decrying Jahn's work. "We have had commentary on our program from no less than six Nobel laureates," Jahn told a gathering of parapsychologists at Fairleigh-Dickinson University in 1983, "two of whom categorically rejected the topic, two of whom encouraged us to push on, and two of whom were evasively equivocal." In "Anomalies: Analysis and Aesthetics" (1989), Jahn describes a typical finding from his lab. While it is expressed in the somewhat bleak, even forbidding, language of science, it's worth quoting to get the flavor of this "high technologist" who has risked his reputation in seeking evidence of the almost unbelievable. Jahn cited a graph, like that in figure 1-1 on page 27, that displays one set of accumulated REG results as three wiggling lines, all starting at the left margin at value zero, and diverging thereafter, still wiggling up and down, until they form a kind of delta or pitchfork stretching to the right-hand side of the page. "The particular case shown," Jahn comments, "pertains to the interaction of one human operator with a microelectronic random event generator

(REG) in a very carefully controlled sequence of experiments extending over nine years."

The graph traces the accumulated deviations from chance expectation of the REG machine's output under the three different conditions described earlier. "The operator alternately attempted to achieve a high number of counts (HI), a low number of counts (LO), or the chance number of counts (BL), interspersed in a random sequence of efforts, with all other technical and procedural aspects of the experiment held identical." In other words—perhaps it's necessary to remind ourselves of how strange this activity is—somebody sat in a room watching a machine churning out random numbers that flipped back and forth in a way quite inscrutable to normal inference and tried to drive the numbers up or down, or hold them in the central range, *purely by thinking about it.*

The result? According to conventional physics and information science, the three separate conditions must produce indistinguishable outcomes—wavy lines jittering and crossing one another around the straight line running through the middle of the graph marking the mean chance value. After all, *willing* a random number generator to go high or low can't have any effect on it—can it?

The graph reveals otherwise. For this person, code-named Operator 10, "the null-intention or baseline effort yields a string of data oscillating stochastically about the theoretical chance mean." *Stochastic* essentially means *at random,* so this BL, or baseline, result is exactly what any skeptic would expect. By contrast, however, "high intention efforts produce results displaying the same sort of stochastic oscillations, but now superimposed on a systematic trend toward ever increasing excess above chance." In other words, the jiggling of chance inputs was being deformed in a skew apparently caused by nothing other than human intention—a skew that carries the cumulative deviations away from the chance line farther and farther in the HI direction on the graph. "The low-intention efforts," Jahn goes on, "show a similar, but even more substantial trend in the opposite direction." Thus, the jiggling line that tracks this set of interspersed accumulating deviations is headed even more markedly in the LO direction.

The impact of these results may not be immediately obvious. Undeniably, the overriding contribution remains the random noise of the background, which accounts for all but two parts in a thousand of the outcomes. But Jahn and Dunne's experiment has done something wonderful—it picked up and concentrated into visibility that 0.2 percent signal from out of the midst of the noise. Nor is this just a trick of choosing the right coordinates on a graph. Statistical analysis bears out Operator 10's paranormal prowess. "For the more than 30,000,000 bits processed in the more than 50,000 tripolar trials of this operator's program," the report concludes, "the likelihood of obtaining the displayed split of the HI and LO data by chance is less than a few parts per million." In other words, to get a single duplication of this result just by chance, you would need to repeat the Princeton operator's years of accumulated efforts in many hundreds of thousands of similar experimental series.

Nor was Operator 10 the only one who could perform these statistical miracles. The really intriguing news was that PEAR found similar effects emerging from the trials of many of their unpaid, anonymous subjects. "More than 30 other operators have performed this same experiment," the report observes. "Some achieve much like [Operator 10]; some are successful in only one direction of effort, or in the other; some display only chance results; a few achieve extra-chance results in directions opposite to their intentions. But despite these major differences in detail, in most cases each operator's pattern is serially consistent with itself, i.e., internally replicable in the statistical sense, so much so that we refer to the individual cumulative deviation graphs as operator 'signatures.'"

These data are explored in much greater detail than can be given here in Jahn and Dunne's 1987 book, *Margins of Reality*, where the outcome, grotesquely unbelievable to ideologically correct scientists, is summarized without embarrassment:

> While small segments of these results might reasonably be discounted as falling too close to chance behavior or displaying yields too marginal to justify revision of prevailing technical or scientific tenets, taken in concert the entire ensemble establishes an incontrovertible aberration of substantial proportions. The case can equally well rest on the overall data bases of certain individual operators. . . . By any reasonable statistical criteria, the likelihood of this concatenation of results occurring by chance is infinitesimally small.

If corroborated, the PEAR researchers knew, it meant that some people at least, under favorable conditions, could modify the effects of microelectronic systems, interfere with random number procedures, and obtain information from the future. The extent to which they did so was absurdly small, but the effect was not zero. Amplified in very large databases, it was highly significant. The gates of science would be forced open by evidence, however unpalatable.

Then something terrible happened.

☙❧

PEAR's encouraging results did not continue unabated. A frequent complaint by critics has been that the immense quantity of cumulatively significant PEAR data, while difficult to attribute to machine flaws or frank fraud (although both attributions have been attempted), is the outcome of work at a single institution. The hallmark of science is replication, which does not mean repeating the same experiment in the same laboratory; other investigators need to run similar experiments, probing for failure points in the original reports.

In 1996, with the cooperation and involvement of PEAR, an international consortium attempted just such a replication of mind/machine interaction. As well as Princeton, the group included researchers at two German institutions: the Freiburg Anomalous Mind/Machine Interactions group (FAMMI) at the Institut für Grenzgebiete der Psychologie und Psychohygiene (the IGPP, or Institute for Frontier Areas in Psychology and Psychohygiene) in Freiburg, and the Giessen Anomalies Research Project (GARP) in the Center for Psychobiology and Behavioral Medicine at Justus-Liebig-Universität Giessen. The result was disheartening, even crushing. In 2000 a joint report acknowledged that although noise-source equipment identical to the PEAR devices was used throughout, with essentially similar protocols and data analysis procedures, "The primary result of this replication effort was that whereas the overall HI–LO mean separations proceeded in the intended direction at all three laboratories, the overall sizes of these deviations failed by an order of magnitude to attain that of the prior experiments, or to achieve any persuasive level of statistical significance." Note that this is not as utterly damaging an outcome as one might imagine. The separations of data correlated with intention were consistent with the earlier findings, just considerably shrunken. That, of course, reduced the replication value of the experiment.

Failure to attain statistical significance seemed definitive to skeptics, given the huge number of events in the attempted replication— each laboratory agreed to provide 250 sessions of 1,000 trials in each of the three standard conditions, each trial being judged by the majority vote of 200 random samples, a total of 750,000 trials and 150 million bits. However, the experimenters added, the brief of the study also required them to extract whatever structural aspects of the data befit its capabilities—including gender effects, serial position effects, standard deviations, feedback correlations, experimenter effects, and other aspects. What they found was that various portions of the data displayed a substantial number of interior structural anomalies in such features as a reduction in trial-level standard deviations; irregular series-position patterns; and differential dependencies on various secondary parameters, such as feedback type or experimental run length, to a composite extent well beyond chance expectation. The change from the systematic, intention-correlated mean shifts found in the prior studies, to this polyglot pattern of structural distortions, testifies to inadequate understanding of the basic phenomena involved and suggests a need for more sophisticated experiments and theoretical models for their further elucidation. (Jahan et al., 2000, 499)

That is, the experimenters concluded that "these replication studies presented instead a substantial pattern of structural anomalies related to various secondary parameters, to a degree well beyond chance expectation and totally absent from the calibration data. To borrow a fluid mechanical metaphor, it is as if the influence of operator intention now was manifesting itself as a structural 'turbulence' in the output data of the replication, rather than in a more orderly displacement of the data streams as was found in the prior PEAR studies" (538). But was this just a case of turning lemons into extremely watered-down lemonade?

The famous Austrian philosopher Ludwig Wittgenstein once asked a student, "Why did people believe the sun went around the earth?"

"Well," the student stammered, "I imagine it was because it *looks* as if it does."

"Ah," said Wittgenstein. "What would it look like if the earth went round the sun?"

This is startling and funny, because, of course, the earth actually *does* go around the sun, and always has, even in the olden days when people thought otherwise. The evidence is unchanged, although these days we do have more of it, such as photographs taken by spacecraft fired far out into space. We can apply the same entertaining yet searching test to other vexed topics: "What would life look like if it had evolved by natural selection?" Or: "How would living creatures look if they had paranormal abilities?"

Science fiction in both its print and movie forms has been telling us for more than half a century that telepaths and their paranormal kin will be mutants, probably produced by nuclear war or the radioactive fallout from bomb tests or nuclear reactors in meltdown, or more recently from strange chemical spills or even genetic engineering experiments gone horribly wrong. These mutations walking among us might be characteristically bald, or have eerie tendrils growing from their domed scalps. Perhaps they'll have two hearts, glowing eyes, the capacity to hypnotize mortals and read their minds with ease or, at worst, suffer a certain discomfort at the brutality they encounter inside our heads. They'll levitate or teleport themselves or other objects at will, command strange energy resources, and gaze into the future or the deep past. Probably they will speak to the dead, and perhaps return them to life—of a kind.

Does anyone take this kind of comic-strip absurdity seriously? I wouldn't be surprised. A recent study of the communal impact of mass-media portrayals of cloning is rather disheartening. In August 2006, researchers at Biotechnology Australia released a study called *Cloning Goes to the Movies* that found Hollywood films are the major source of information for young people on human cloning, resulting in some bizarre ideas. "One of the misconceptions is that it's like running a photocopier across an animal or a person, and bing! There's a life-size person, a perfect double." You can never be sure that the people answering such polls are not pulling the researcher's leg, rolling their eyes as they do it. Still, if the survey is valid and "many people believe even the most far-fetched movies about human reproductive cloning are based on fact," the results are likely to apply to the same kinds of folk beliefs about psychic abilities.

Leaving aside the endless horror movies, both frighteners and spoofs, television has lately presented a range of good-looking young women whose paranormal powers verge on the supernatural: Buffy the Vampire Slayer, of course, and her attendant witches, zombies, demons; the Halliwell Charmed Ones; and more recently a pair of mediums who intervene to the benefit of the living with the shades of the dead. Extraordinarily, the CBS television series *Medium* presents an engaging paralegal and her aerospace scientist husband, together with their small children; the woman character, played by Patricia Arquette, is given the name "Allison DuBois"—which, in reality, is the name of a Phoenix medium in her early thirties, married to an aerospace scientist, with three children, and, judging by the photographs, as attractive as the good-looking actor playing her in the series. While DuBois has taken pains to stress that the events of the program are fictional, the contrivance of clever scriptwriters, the validity of her mediumship has been attested by Dr. Gary E. Schwartz, a professor at the University of Arizona, who has already published a small volume dealing with her purported gifts. So we have art imitating life even as the last of the goth paranormal devotees continue to parade in black as they imitate art.

None of these portrayals of the paranormal, the psychic, and the parapsychological come especially close to the real world. One might hope that the zanies in the 1984 movie *Ghostbusters,* with their electronic detectors, fizzing containment traps for catching and sequestering disincarnate horrors, and proton-pack energy streamers, took their inspiration from real parapsychologists, but alas, no. Or perhaps only in their dreams. Although I fancy to detect a resemblance between actor Harold Ramis's brilliant ghostbusting character, Dr.

Egon Spengler, and Dr. Rex G. Stanford, 2006–2007 president of the Parapsychological Association, the scholarly body affiliated with the AAAS. Dr. Stanford, who with his twin brother, Ray, channeled extraterrestrial messages in a mediumistic trance when he was a Texan teenager but later repudiated it all as delusional, teaches cognitive science at St. John's University, a Catholic University in New York. In 1974 Stanford advanced a remarkable model of psychic functioning that places psi in the context of ordinary life, which for most organisms, including humans, is the contestatory environment governed by evolution: the striving to survive, thrive, and reproduce successfully. Stanford called it the PMIR model, for Psi-Mediated Instrumental Response, which is unfortunately unpronounceable and not nearly as catchy as "ESP."

In a 1990 update revision, Stanford declared: "[E]volutionary considerations would seem to militate against the idea that extrasensory response comes into play *only* when a person consciously wishes to gain knowledge of something outside the reach of the senses. There are many life situations in which psi-mediated adaptive response to information outside one's sensory knowledge would be vitally important, yet the individual might not even know that such information exists (e.g., the individual may have no sensory knowledge that a predatory animal is approaching and may not even be thinking about the possibility)" (55). By contrast, the participant in a parapsychology laboratory usually knows in advance exactly what kind of task he or she will face, although, naturally, the details are not knowable by normal means. Hubert Pearce had to guess his way through a pack of Zener cards; he knew when to start guessing; he knew what the five options were. These days, as we'll see, a remote viewer probably has far less notion of the kinds of pictures, people, or places he or she will be asked to identify to psychic means, but at least there is no major surprise involved. Wild dogs are not going to be unleashed suddenly inside the laboratory, menacing the psychic or the psychic's children.

So it is arguable that although psi can indeed be detected under contrived, artificial laboratory constraints, that is not the environment it was evolved to probe and defend. This assumes, of course, that psi—like sight and hearing and the capacity to detect the stench of an ant when you step on it—is in fact an evolved function, rather than, say, a side effect of an immaterial human soul coupling to the gross matter of a human body but leaking information at the edges.

One of the key discoveries about living creatures and their behavior, illuminated pitilessly by evolutionary theory, is how messy and undesigned organisms really are on close inspection, however sublime they seem to us. Because natural selection is incapable of looking ahead, but can only retain the accumulated successes of the past, including some that have just happened by accident in the latest batch of genomic mutations, species improve by the gruesome process of deletion. Benefits accumulate, certainly, but they do so by a process that computer specialists call a *kludge* and anthropologists a *bricolage:* an ad hoc adaptation, built from the bits and pieces that happen to be lying around on the workplace. If psi is an evolved function or capacity, we shouldn't expect it to be perfectly honed, any more than we expect old people to have perfect eyesight or excellent hearing. Evolution is blind to the elderly's encroaching blindness, because their genetic constitution has become invisible. Their health and endurance is no longer a factor directly detectable by evolution's winnowing comb. They produce no new offspring (although the survival of elderly wisdom has its merits), so in general a mutation that aided the old probably would have no survival value in the young. Indeed, most likely it would prove costly and wasteful of energy in the years when the young are growing in strength, making their way in a sometimes hostile world, and rearing their own young. Evolution, in other words, does not optimize to the benefit of its players. We need to borrow peculiar words from economics and game theory, words like "satisfice" that combines "satisfy" and "suffice." Psi is probably like that, too, evolved in the usual catch-as-catch-can fashion, retained for its marginal benefits, reinforced by its marginal utility. If you approach the phenomena with the eager devotion of a spiritualist or a mystic, expecting revelations from on high or from beyond the grave, you might well find them—but whatever you find, it is unlikely to bear much resemblance to the effects of a prowess that has evolved, like every other, primarily to help you survive and pass on your genes in a dangerous world.

Dr. Richard Broughton, a past president of the Parapsychological Association, raised these issues in his 1987 presidential address:

[Q]uestions concerning why we have psi or its role in life have traditionally been awkward to deal with. . . . If we are going to take psi seriously as an ability, then we have to have a better idea of its function in real life before we try to capture it in the laboratory. . . . The real point to consider, and one we must keep in mind when trying to study psi in artificial situations, is that probably the primary function of psi is to help individuals survive. . . . *and also to gain a competitive advantage in the struggle for survival.* . . . It could well be elusive and obscure by design. . . . The most effective psi might be imperceptible psi. . . . If psi is designed to serve important needs, then the way to test for psi is in situations where the use of psi will have *consequences* for the individual. (196)

It would be depressing if it turned out that Joseph Banks Rhine, for all his immense influence in having the phenomena taken seriously, had led the scientific study of the paranormal down a blind alley for fifty years, leaving it stranded outside the gates of science. Luckily, around the start of the 1970s, the quest for psi, for its applications in the real world, and for its explanation, began again in earnest. Almost unbelievably, the main impetus came from American researchers in programs sponsored and funded by the CIA and the U.S. armed forces, programs that were classified until 1995. Since then, there's been a slow leakage of information and a torrent of disinformation, duplicitous exploitation, crazed abuse leading to mass suicide, with most of what is valuable entirely ignored by the press and the media. Let's watch this strange saga unfold.

2.
THE HALF-TRUTHS ARE OUT THERE

The 1970s saw the emergence of an astonishing psychic phenomenon that the principal investigators called "remote viewing." This refers to an alleged ability to perceive information from remote sources not available to any known sense. Actually this product of the paranormal is not a new one—it is really good old faithful ESP in a new package and with a different brand name.... The results were allegedly reliable and repeatable. In fact, remote viewing was every parapsychologist's dream come true.... [I]t became clear that the SRI research program, promoted as well-controlled science, was actually a massive litany of fallacies and flaws.

> —David F. Marks, "Remote Viewing Revisited"

For the larger part of the scientific community, the first anyone heard of remote viewing was a 1974 peer-reviewed article (or "letter") by Russell Targ and Dr. Hal Puthoff in the august pages of *Nature*, one of the world's major science journals. Since the authors described nothing less than information acquisition through the medium of psi, this was an extraordinary event in itself. Perhaps nothing so offensive to scientific orthodoxy would be presented until the scandalous

announcement, and instant establishment debunking, of "cold fusion"
by Drs. Stanley Pons and Martin Fleischman in 1989. That first paper
on remote viewing, "Information Transfer under Conditions of Sensory Shielding," was followed up in 1976 by a more complex account,
from the same authors, in the equally reputable *Proceedings of the
IEEE* (Institute of Electrical and Electronics Engineers, Inc.). What
few knew at the time is that the experiments reported in these documents had led to a well-funded, classified government investment
into psychic phenomena, tasked to provide ancillary intelligence information to various branches of the military and intelligence services,
such as the CIA, and to pursue scientific investigation into the nature
and possible applications of psi.

Puthoff and Targ, the two laser scientists who initiated this quest
into the paranormal, worked for a large private Californian think
tank that frequently did classified work: SRI, short for Stanford
Research Institute, in Menlo Park (later SRI International). These
days Puthoff is a private researcher working in Austin, Texas, on an
off-center theory of physics known as quantum electrodynamics,
which explores the ubiquitous quantum zero-point field. A former
naval officer, he graduated from Stanford University in 1967, and his
résumé lists General Electric, Sperry, the National Security Agency,
and the CIA; he holds a Department of Defense Certificate of Commendation for Outstanding Performance. Targ received a bachelor of
science degree in physics from Queens College, New York, in 1954,
with graduate work at Columbia University, before developing lasers
for Sperry, GTE Sylvania, Lockheed, and SRI. Unlike Puthoff, Targ
remains directly involved with psi research, especially in the area of
healing by beneficent-directed intention. By a brutal irony, his
daughter Elisabeth, a psychiatrist, died in 2002 from the aggressive
brain tumor glioblastoma multiforme while working on psychic
healing methods for curing just that disease.[1] Puthoff and Targ developed an interest in the possibility of utilizing psi when they heard

about the alleged prowess of a New York artist and psychic, Ingo Swann, invited him to California in 1972 for controlled testing, and famously observed his capacity to modify the state of a highly protected and stabilized magnetometer meant to register only subtle activity at the subnuclear level—and then to perceive and describe with some accuracy its hidden internal workings.

In an online memoir of the remote viewing saga, Swann comments:

> I would like to emphasize that if Hal Puthoff and I had not gotten together, it is COMPLETELY likely that there would be no story to tell. I will next frankly state that without Puthoff, remote viewing would never have come into functional existence and the high visibility it did. I may have been the lean, mean, fighting machine (as many were to discover, including Puthoff). But Puthoff was the diplomat-warrior who held the enormously complicated remote viewing "package" together—a feat I could not have pulled off myself.[2]

After Puthoff circulated a brief note on his initial findings with Swann, some spooky attention was captured. The intelligence community was already concerned at rumors of psychic breakthroughs behind what was then the Iron Curtain. Puthoff recalled:

> In a few short weeks a pair of visitors showed up at SRI with the above report in hand. Their credentials showed them to be from the CIA. They knew of my previous background as a Naval Intelligence Officer and then civilian employee at the National Security Agency (NSA) several years earlier, and felt they could discuss their concerns with me openly. There was, they told me, increasing concern in the intelligence community about the level of effort in Soviet parapsychology being

funded by the Soviet security services; by Western scientific standards the field was considered nonsense by most working scientists. As a result they had been on the lookout for a research laboratory outside of academia that could handle a quiet, low-profile classified investigation, and SRI appeared to fit the bill.

Targ and Puthoff's research, funded by these and other visitors, grew to embrace other psychics, including former police commissioner Pat Price, whose remarkable feats of remote viewing were described in *Nature,* alongside some carefully constrained accounts of research with Uri Geller, who by that point had already become a media sensation and an embarrassment to anomalies researchers. Then-editor of *Nature,* Dr. David Davies, introduced this controversial October 18, 1974, issue with a substantial editorial, "Investigating the Paranormal" (559–60), which took pains to point out that Uri Geller's rumored abilities to bend metal rods by stroking them, and all the rest of the shenanigans, were not addressed in a paper that did, admittedly, proffer the claim that "some people can read thoughts or see things remotely" (559). It's worth considering Davies's extensive apologia, to gain a feeling for how the community of scientists regarded this topic more than three decades ago and then contrast that with the way scientists feel about psi these days. The atmosphere seems much bleaker today. I expect that this will change, that psi will be invited in from the cold. But, then, Targ and Puthoff seemed to have pretty much the same expectation back in 1974.

Davies wrote judiciously: "The publication of the paper would be justified on the grounds of allowing scientists the opportunity to discriminate between the cautious, limited and still highly debatable experimental data, and extravagant rumor" (559). The editor notes that he had "corresponded informally with the authors on one or two issues such as whether the targets could have been forced by standard magical tricks, and is convinced this is not the case" (560). He stresses that publishing in the scientific journal is not in itself "a process of receiving a seal of approval from the establishment; rather it is the serving of notice on the community that there is something worthy of their attention and scrutiny. . . . Perhaps the most important issue raised by the circumstances surrounding the publication of this paper is whether science has yet developed the competence to confront claims of the paranormal" (560).

Just such protocols were being developed under the aegis of the American government, which lends a certain irony to Davies's cautionary comment: "Many scientists and very large numbers of nonscientists believe, as a result of anecdote or hearsay, that the Stanford Research Institute (SRI) was engaged in a major research program into parapsychological matters and had even been the scene of a remarkable breakthrough in this field" (559). In light of the material declassified some twenty years later, and the testimony of participants—scientists and psychics alike—it seems clear in retrospect that this is in fact precisely what was going on. Whether "remarkable breakthroughs" were achieved remains a matter of some dispute.

To parapsychology insiders, not much was new. The heft of this letter to their fellow scientists was its publication in such a notable, peer-reviewed journal. Of course, anomalies research has its own peer-reviewed outlets, but it is no accident that conventional scientists regard these journals with suspicion, as nothing better than coreligionists preaching to one another.

Two of the three segments were pretty much old hat. In the first, Uri Geller was put through his paces under what appears to be fairly stringent controlled conditions (inside a visually, acoustically, and electrically shielded room), drawing simple sketches of hidden target images previously prepared by SRI staff. This is the sort of "mental radio" experiment conducted in the 1920s by the celebrated novelist Upton Sinclair and his wife, Mary Craig, and reported in the 1930 book of that title. What's interesting about Geller's performance, as illustrated in figure 1 of the paper (603) is how often his images, which do bear a striking structural if sometimes misguided resemblance to the original, transpose left and right, or even include fragments of imagery from adjacent trials (a devil with a pitchfork seems to contaminate another image showing the earth with what looks like the two tablets of Moses). There are two ways to interpret this occasional transposition or mirror reflection: It might be a magicians' trick to throw the gullible off track, to toss their credulity a bone, in fact, by making a small but forgivable mistake, thereby "proving" that the magician hasn't cheated. Alternatively, it might reveal something extremely significant about the way the mind/brain transduces into conscious awareness imagery received by psi. We know, for example, that the retina of the eye receives an upside-down image of light coming to us from the world after passing through a lens that inverts the image. In some cases, upside down is exactly what happens also with remote viewing representations of the scene, leading one to wonder if the psychic image has been captured by some deep internal part of the brain's visual system without having gone through the preprocessing at the immediate post-retinal stage. That would not explain mirror reversal, but perhaps some equivalent neurological quirk comes into play when non-sensory conduits bring us information. Or, again, maybe it's just a cunning trick played by someone intent on deceiving us.

Remote viewing in the now-classic sense was described in the second section of the *Nature* report. This method was used by other psychic researchers, notably Brenda Dunne, later to play her important role at PEAR as an operator and experimenter. Here we find rather austere descriptions of a series of remote perambulations by Pat Price, or perhaps his thetan (a cult term we'll get back to). In nine attempts to describe certain locations that were a thirty-minute drive from the laboratory, each drawn at random from a set of "natural geographical places or man-made sites that have existed for a long time" on the grounds that these are "more potent targets for paranormal perception experiments than artificial targets prepared in the laboratory," Price did not simply say, "Hmm, this looks like a mall to me." He correctly described "buildings, docks, roads, gardens . . . structural materials, color, ambience and activity, sometimes in great detail." The descriptions, Targ and Puthoff were careful to point out, "contained inaccuracies as well as correct statements" (605).

In order to evaluate how well Price's fantasizing and sketches fitted the actual target places, five SRI scientists otherwise unconnected with the project were dispatched to the nine locations with typed transcripts of the tape recorder narratives. All five correctly identified the Hoover Tower from its transcript. Four identified the Redwood City Arena when that was the target. Four of them correctly picked an Arts and Crafts Garden Plaza. In the remaining transcripts, each chosen target was selected from its rivals by a plurality of the judges. Using a highly conservative judging evaluation that scrapped all the subtle detail in Price's remote viewing records, the probability of getting a result this good just by chance was determined to be less than one in one thousand. Not bad for a strictly controlled experiment—or was it?—that was intended to constitute a first step toward the goal of uncovering "patterns of cause-effect relationships that lend themselves to analysis and hypothesis in forms with which we are familiar in scientific study" (602).

The report's third section was the most innovative, but also the least successful. If sometimes we can make use of psychic information arriving principally at an unconscious level, subject to noise and false interpretation from conscious processes, might we devise an end run around interfering minds and see if we can detect activation of parts of the brain's functioning to which we have no direct access? This is an approach that continues to this day, using ever more subtle devices (functional magnetic resonance imaging scans, for example). Targ and

Puthoff examined the electroencephalogram (EEG) output of subjects ("receivers") sealed into the same shielded, inaccessible room where Geller had done his clairvoyance. Meanwhile, in a distant room, the eyes of a sender were stressed from time to time with a flashing light, driving the sender's EEG into the same frequency as the flashes. Interesting question: would the brain of the "receiver" fall into resonance with the distant sender's EEG? Actually, no. In control runs, of course, the receivers did show photic driving when their own eyes were subjected to flashing. The most capable remote viewer in the group, Hella Hammid, did show alpha blocking when the sender was watching flashes. Slow alpha waves typically vanish when the eyes are open and looking at something. In a series of follow-ups, this appeared to be confirmed.

Target and Puthoff offered three conclusions, couched in terms appropriate to physicists and engineers, that implied a theory of communication in which these explicitly extrasensory sources of information are construed as similar to ordinary sensory channels and processes:

> A channel exists whereby information about a remote loca-
> tion can be obtained by means of an as yet unidentified
> perceptual modality. As with all biological systems, the
> information channel appears to be imperfect, containing
> noise along with the signal.
>
> While a qualitative signal-to-noise ratio in the
> information-theoretical sense cannot as yet be determined,
> the results of our experiments indicate that the functioning
> is at the level of useful information transfer. (607)

There's a hint here, that we can now see in hindsight, of attempts to *use* psi for practical applications, to put it to work in the service of their sponsors. At that time, though, only Uncle Sam, and at least one spy, knew the details.

Despite the security classification of much of their work, Targ and Puthoff published a considerable amount about their techniques and findings in two popular books: their *Mind-Reach* (1977), and *The Mind Race* (1984), the latter written by Targ with a young psychic and parapsychologist, Keith Harary. Were their intelligence community sponsors angered by these disclosures? Apparently so, and repeatedly. Was the scientific community swayed from its general skepticism by their reports? No, not a bit. In fact, David F. Marks and R. Kammann, two psychologists at a New Zealand university, quickly published a damaging critique, *The Psychology of the Psychic* (1980). Marks savaged Targ and Harary's book in his critical and somewhat persuasive "Remote Viewing Revisited" (1986).

Meanwhile, though, under the protective shadow of security classification, the secret project into remote viewing and other psychic phenomena continued at SRI and several military bases, then moved to SAIC (Science Applications International Corporation), the largest employee-owned research and engineering company in the USA. A considerable amount of documentation was formally declassified in July 1995, following President Clinton's Executive Order 1995-4-17, titled "Classified National Security Information." In November of the same year the lid was blown off publicly on ABC television's *Nightline*, a current affairs program that interviewed a highly decorated former director of the Central Intelligence Agency, Robert M. Gates (in late 2006 named as secretary of defense), and an unnamed CIA operative attached in an oversight role to what was now known as the Star Gate program. The interviewers also spoke with Dr. Edwin May, longtime director of the scientific program, and, in background comments, Legion of Merit holder and military remote viewer Joseph W. McMoneagle. The testimony of May and McMoneagle made it clear that serious discoveries had been made and serious results had been obtained; the CIA men metaphorically shuffled their feet, bit their lips, and pretended unconvincingly that nothing had happened worth paying any attention to. Inside a year, though, the Star Gate program had been shut down, with assessment documents released to reinforce the official message to the media, a message swallowed with gusto: the whole thing had been a big mistake, nothing to see here, folks, move right along.

All of this history, public and covert, is slowly coming to light, and much of it is already available on the Internet. But like everything else on the Internet, the truth is corroded, corrupted, and swamped by cascades of self-serving lies and opportunistic half-truths by people once marginally connected to the program, by muddle, mystical claptrap, and probably official disinformation. Puthoff has published a clear and interesting brief account of the program, "CIA-Initiated Remote Viewing at Stanford Research Institute." He notes that "although I had founded the program in early 1972, and had acted as its Director until I left in 1985 to head up the Institute for Advanced Studies at Austin (at which point my colleague Ed May assumed responsibility as Director), it was not until 1995 that I found myself for the first time able to utter in a single sentence the connected acronyms CIA/SRI/RV." In other words, despite a substantial amount of information already released by these researchers, and despite scuttlebutt in the press by such insider reporters as Jack Anderson, nobody until 1995 officially admitted that the government and the military had paid out $20 million for this kind of research over more than two decades.

After all these years of watching vivid conspiracy-based movies and television programs like Fox television's *The X-Files,* the shock of the government's disclosure is probably long since gone. Still, knowing that this sort of research has been going on *in reality,* rather than just in entertainment fiction, must surely be a jolt to the system. How much more surprising would it be, to the average skeptical scientist, to learn that the CIA and other spooky entities had been funding Santa Claus research, or even, let's say, reincarnation or astrology research, or claims that time-traveling intelligent dinosaurs from the remote past were visiting us in UFOs? On second thought, none of these latter possibilities is *entirely* out of the question. In a way, it might be refreshing to learn that what we mostly regard as hidebound bureaucratic structures harbor within them small, secluded, protected enclaves of imaginative craziness and inspired silliness. At any rate, I'm pretty sure I do understand at least part of the mysterious process that led to the formation of psychic studies supported by the intelligence community. It resembles the way traditional psychic research was kick-started at the end of the nineteenth-century by scientists who were intrigued by spiritualist claims. A blend of hope and non-conformist ideology impelled them to look for evidence that mind was more than just body and brain in action, that an indwelling soul or spirit might persist beyond death and return when asked nicely to speak to the living. In the 1970s spiritualists were thin on the ground—despite the efforts of Shirley MacLaine—but plenty of cults, gurus, and "training systems" were available for would-be believers to embrace, hoping that life exceeded the literally mindless reductionism of behavioral psychology that even then was on the way out, kicking and clawing as it went.

One of these cults, which had a surprising success given the way it began and continues to this day, was Scientology. Its founder, L. Ron Hubbard, had been a charismatic adventure-story writer with a penchant for tall tales and ceremonial sex magic who took up science fiction fantasy in the 1930s and 1940s, pouring out stories for pulp magazines at a quite extraordinary rate, before in 1950 announcing to the readers of John W. Campbell's *Astounding Science Fiction* magazine that he had discovered the secrets of how to engineer the mind, remove the sources of neurosis and any physical diseases derived from mental problems, and develop the power of a cleared mind. This new doctrine of mental health, originally called Dianetics, swiftly became fragmented but was later salvaged in the guise of a religion, Scientology, that operated like a Masonry or pyramid scheme for the soul, with many embedded esoteric levels of seniority, levels that increased in number—and expense and commitment to the initiate—as Hubbard's organization expanded and grew in ambition. It fell upon hard times in the 1980s, after Hubbard's wife, Mary Sue, was arrested, along with some of the top apparatus, and jailed for tax evasion and audacious thefts of confidential information from elements of the federal government.

By the twenty-first century, however, with Hubbard dead, his son Ron Junior having denounced the organization, and another son dead by his own hand, the organization recovered by appealing to such media stars as, famously or infamously, Tom Cruise. As a 2006 report in Britain's *Guardian* newspaper put it rather deftly (since Scientology has been from the outset extremely litigious, deterring criticism or even ridicule): "[Cruise] demanded that a Scientology tent be erected beside the movie set of *War of the Worlds*. And he shouted down the U.S. television host Matt Lauer for daring to suggest that pharmaceuticals might sometimes help treat psychological disorders. . . . Scientology views itself as waging a war with psychiatry, arguing instead that the root of human misery lies in the actions of an intergalactic ruler who arrived on earth 75 million years ago with a fleet of spaceships."

One of the key doctrines of this pulp science-fictional religion is that our "meat bodies" are hosts to extraterrestrial, indeed extragalactic, nonmaterial entities called *thetans*. It would be distracting, however entertaining, to explore the multitrillion-year history of the cosmos according to Hubbard, but note that with some training the thetan—the core and consciousness of the person—can be "exteriorized," and even leave the room and jaunt around the planet, returning to the meat body with memories of its peregrinations. This is pretty much the same thing described as "out-of-body experience," or "traveling clairvoyance," or, indeed, as it came to be known, "remote viewing." Perhaps it is no coincidence, then, that when Puthoff started the program, he was a scientologist of high degree, as was Ingo Swann, as was Pat Price, as was, for a time, Edwin May, as, indeed, were a number of the military connections to the program that eventually became known as Star Gate.

Does this disclosure imply anything *dubious* about the history of the remote viewing program? Not necessarily—any more than does the adherence of a trained professional biologist to fundamentalist Christianity, although one might not be entirely surprised if the biologist's affiliation led to her embracing the bogus pseudoscience of "intelligent design." Dr. Francis Collins, longtime director of the U.S. government's Human Genome Project, is a devout Christian; this has not interfered with his ability to decode the human genetic blueprint. The early-twentieth-century scientist who first came up with the big bang theory of the origin of the universe (although not under that name) was a Catholic priest, Abbé Georges-Henri Lemaître, and if his devotions encouraged him to believe in a creation moment, science has embraced that model only because it fits the evidence so well.

One egregious psychic scandal, though, has been revealed by CIA physicist and analyst Dr. Kenneth A. Kress. Pat Price, an "Operating Thetan Level IV," repeatedly debriefed his Scientology auditor (someone like a Catholic priest hearing confession, or a psychoanalyst listening to her patient's dreams and free associations, but without the seal of the confessional or the couch), providing information and stolen documents on SRI's early remote viewing program and contacts—breaching his sworn oath in doing so (Kress, 1999). Kress writes:

In the late 1970s . . . I got a secure line call from a person who identified himself as an FBI agent. He suggested that I should be prepared for a spate of publicity about the remote viewings of Pat Price . . .

The FBI agent proceeded to explain that Pat Price was a member of an organization that was recently raided for documents indicative of illegal activity. The organization was vigorously resisting the government investigation but the raid produced hundreds of files and papers that supported the government's allegations. These documents were now in the public domain as part of the discovery process in the legal proceedings.

One such file included debriefings of Pat Price about his CIA remote viewing projects. The debriefings were a detailed record of the intelligence objectives I had given Pat and results that Pat provided to me. The files revealed the meeting places as well as all the names of those present. My esteem instantly rose for my colleagues who had used first names only with all meetings with Pat! As the file made clear, Pat, who had signed an official secrecy agreement, would immediately go to his superior in the organization after sessions with me and divulge everything. As far as I know, the documents were never read by anybody who publicized them and the organization never used them.

Shades of the Cold War spies and traitors of the mid-century! On the other hand, the principal CIA psi players are said to have abandoned L. Ron Hubbard's comedic cosmology a good many years ago, assuming they ever knew about it; apparently those not on the top rungs of the cult are shielded from these great galactic truths, although Puthoff is said to have attained Level III Operating Thetan status, and Ingo Swann the even loftier OT Level VII. Perhaps we can put the Scientology connection out of our minds, except as a piquant reminder of the occasional spark-plug link between strange beliefs and the embrace of strange or radical science. (One further doubt might linger: is it possible that with so many key players committed to a zany belief system, some might have colluded in contriving the evidence for remote viewing? The best rebuttal can be found in equally impressive and independent work done both previously and subsequently in other research laboratories, sometimes by avowed atheists, and probably not much of it contributed by cultists—although it must be admitted that many of the researchers in this area appear to be motivated by spiritual longings and convictions.)

So much blood, sweat, and tears have flowed under the bridge in the last three-plus decades that there's not much point in wading back into it. A quick overflight might be worthwhile. After the public dismantling of Star Gate, several intriguing books have offered incomplete accounts of the kinds of protocols developed by the research program, but Targ and Harary's popular 1984 book, *The Mind Race*, remains a convenient aperture into the early days of remote viewing. Without breaching the bounds of classified information, Targ described pioneer work with Price, Hammid, and other psychics. To get the full impact of these reports, you really need to ponder the transcripts, look at the sketches, and then compare photographs and descriptions of the actual targets. Only a faint flavor of the thing can be conveyed in summary. For example, photographer Hella Hammid, sitting with Targ in a windowless experimental room on the third floor of the Radio Physics building, tried to remote-view the scene visited by an SRI "outbound" or "beacon" person. The target, not knowable by ordinary means by either Hammid or Targ, and never used previously as a target site, was one of sixty filed possibilities. It was identified thus to guide the beacon driver: "AIRPORT TOWER—Cross Bayshore Freeway on Embarcadaro Road. On your left will be the Palo Alto Airport. Follow the Airport road to its end, which will take you to the tower. Notice the square concrete base, and pentagonal glass top. Walk around the tower, look up at it. Also notice the aircraft nearby, and the surrounding trees."

One of Hammid's drawings is displayed in *The Mind Race,* alongside a photograph of the site. Seeing one below the other, there is no doubting the resemblance—but it's not clear to me that a "blinded" judge would necessarily identify one from the other, except by contrast with several other decoy possibilities. (That compare-and-rank procedure was used in this experiment, but in a psi application trawling for a missing person, for example, no one would have that advantage of trying for a best fit from a very restricted number of possible targets known to the judges.) If anything, the diagram seems to show a trilothon, two great standing stones crossed by a sarsen lintel, with huge pointed leaves rising at forty-five degrees from the base. Of course, a leafy Stonehenge might not be the kind of spectacle one expects to encounter within a thirty-minute drive of SRI, even in Los Angeles.

Luckily, Hammid's verbal notes were more acute still:

> towerlike building . . . could be bushes around the base of it,
> or flower beds with ornamental designs. And the tower seems
> to be facing a gully, or river, or a stream. . . . It seems to be
> quite square, it is not a round tower, it is a square tower, with
> slightly curved surfaces. But definite corners, four corners,
> and it does seem to have a sort of enlarged tip at the top . . .
> from the top it looks like it has winglike projections on either
> side of it. . . . There is some kind of a cross grid. . . . I get more
> and more the feeling that it is an installation, some technical
> installation. Like a weather station, or an airport tower.

Reading this, in conjunction with the drawing, and assuming that no other decoy target more closely resembled these descriptors, it would be easy to suppose that Hammid was reporting on imagery resembling a memory not quite recovered or pinned down correctly, with the wings of nearby aircraft rendered as pointed leaves, yet with the tower characterized with quite remarkable accuracy. Reasonable skepticism is required, of course, in assessing any such report, but this kind of apparently impossible precision can make the hairs rise on the back of your neck.

What is psychology professor and critic Ray Hyman's response? In a critique of *The Mind Race,* he agrees that "the verbatim transcript and drawing made by the viewer, Hella Hammid, indeed seem to match the target well beyond any forced matching that one can usually achieve between a scenic description and a reasonably complex geographical site" (1986, 100–01). Even Hyman is impressed! Is he *convinced?* Well, no. "But this particular session occurred after three preceding unsuccessful sessions. A skeptic might want to study all the transcripts in this series before jumping to conclusions about possible psychic correspondence" (101).

Sounds fair. If the chance of having the transcript judged first of four possibles was, by definition, one in four, it's not that impressive—especially since it's the one correct match in four sets. Coincidence, then? It seems to me it would be foolish indeed to rest in satisfaction on that analysis. After all, Hammid's correct identification was *loaded* with detail, much of it exactly right, much more structurally correct. The one way I can think of to dismiss such a feat is to suppose that she had a mental repository of perhaps fifty or one hundred notable visual sites within a thirty-minute commute of SRI (she was a professional photographer, after all), and this time she had the one-in-fifty-or-one-hundred-or-so luck to pick the right one—which of course she was then bound to elaborate impressively from memory. Had she chosen the *wrong* target, nobody would have noticed the richly embroidered quality of her errors—it would have been marked down as a miss, and the experimenters would have sighed and gone on to another trial. But that beneficial asymmetry in estimating the weight of a detailed hit only matters if Hammid actually *did* know the area especially well, if perhaps she had been to the Palo Alto airport tower before. I don't know how one can determine the truth of such a conjecture at this point. Alas, we can't ask Hella, as she died in 1991. (After her death, according to remote viewer Jane Katra, PhD, a specialist in public health, Hammid aided Katra in curing Russell Targ of terminal liver cancer. Belief in postmortem persistence seems unusually prevalent among anomalies researchers, perhaps not surprisingly.)

So the interested, open-minded onlooker is stymied, at least to some extent. Reading through the remarkable case studies in a book like *The Mind Race*, complete with photographs of targets and relevant portions of remote viewing transcripts, it's hard not to be dazzled. Keith Harary himself provides examples of precognition experiments where both he and Hammid served as subjects. We see an example drawn from Harary's own trials: a box top from a coin holder that resembled a small round mirror inside a square frame with small holes or circles at each corner, which elicited a picture of a small square mirror in a square frame with small holes or circles at each corner; a stack of silver dollars, inside the box, represented as a cylinder of piled arcs and the word "corrugated"; and a cylinder with the word "metal bands." The two independent judges gave this a 100 percent mark for accuracy. Bear in mind that the target might have been any small object. In another comparable experiment, the array of possible targets included a book, a telephone, an alarm clock, a small doll, two bananas, a camera, and other small objects. It seems gratuitous skepticism in the extreme to suppose that Harary's identification of so many components of this coin holder target is just accidental coincidence.

In fact, Harary had an explanation for the resistance of skeptics. He wrote, bewildered, that "with so much excellent psychic functioning being demonstrated in research laboratories, it is a wonder that most people have not been trying to get into the act." That was in 1984. Within a decade and a half, a very large number of people *were* in on the remote viewing act, making big bucks selling the supposed secrets of psychic functioning. Some of these people were centrally associated with the SRI and other classified experiments, some tangentially associated, and many apparently friends of someone who knew the cousin of someone who had seen it done on television. Anyway, here is Harary's explanation for skeptical resistance: "Critics of psi . . . condemn the scientific work in this field out of fear of its philosophical implications. . . . Other critics are motivated mainly by their own private fear of the unknown" (111, 156). This is possible, of course, but how on earth does Harary know this? Sorry, silly question—he's a psychic, after all.

Or is he?

Today, Keith Harary holds a PhD in psychology and is trying hard to shake off the scandalous and presumably career-damaging associations of earlier days. In essays and interviews, he has taken pains to *deny* that he is a psychic, and (on the face of it) even to question the existence of the phenomena. Of course, he has a perfect right to do this. He was a young man at the time, and young men learn better as they grow older. Still, it is extremely startling to read in *Omni* magazine (April 1994, just ten years after his book came out): "I was a teenager when I first believed the lie that there was something about me or anybody else that could properly be labeled 'psychic.' A part of me felt sick when the label was used on me—the way I felt when I smoked my first cigarette. . . . Reviewing nearly 25 years of experience in psychological research and the findings of more than 100 years of parapsychology experiments, I cannot point to any evidence indicating that humanity can objectively be divided between psychics and nonpsychics."

You can easily gain the impression that he's denying the reality of all those paranormal events he had described so enthusiastically in his book with Targ. Not quite. Harary is making a rather subtle distinction, I think, perhaps so subtle that many will think he's off the psi hook without a mark on him. "If a higher perceptual, communicative, and thinking capability exists within us, then it cannot be destined to remain anomalous or denied by rational people or consigned to the realm of the psychic and paranormal. It must be understood within the context of normal experience and achievable human potential and considered within the emerging framework of mainstream science."

Well, that seems fair enough, doesn't it? Psi should be fetched back inside the gates of authorized science. There are no psychics, Harary is saying, not because there are no *psychics* but because *everyone* is psychic, more or less. It might not seem an *absolutely* plausible claim, since it's a bit like saying there are no Olympic-class athletes, just us several folks here and all God's chillen got wings. That's a very wholesome standpoint, ethically speaking, unless of course you are a physiologist or trainer hoping to discover what makes world-class athletes different from everyone else.

By 2005 Harary was distancing himself even more from his earlier acceptance of psi and propaganda in its favor. In an article titled "Confessions of a Star Psychic" in the November/December 2005 issue of *Psychology Today,* he explained that his youthful anomalous experiences with the American Society for Psychical Research and the Psychical Research Foundation were largely a matter of misconceptions and coincidence. Later, working for Star Gate, he says, he was tasked to locate somebody of interest to the intelligence community. Without using formal remote viewing techniques, Harary reported his impressions of an ill man who would shortly be taken on to a plane: "'One side of his body seems damaged or hurt.' . . . The target turned out to be the hostage Richard Queen, held by Iranian militants and now desperately ill with multiple sclerosis that affected his nerves on one side. In part due to my input, I was later informed by contacts at SRI, President Carter dispatched a plane to bring Queen home."

You can see why the gullible might suppose that Harary is indeed capable of psi remote viewing. Harary, though, it seems, increasingly resisted this interpretation. Although he joined Targ in forming Delphi Associates to make money using precognition by forecasting whether silver prices would rise or fall, and called it correctly nine times in a row before failing twice and terminating the project, this apparently impressive feat is now consigned to the junk heap of dubious science. Harary declines to explain why that should be the move of choice, nor does he mention a failed lawsuit against Targ. Here is his general explanation for apparent psi: "I'm the first to admit that striking correspondences between individual descriptions and targets in various experiments have sometimes seemed incredible. I've seen subjects describe stuffed animals, architectural details and hidden pictures with such fine detail it appeared impossible to describe them better. It seems difficult to fathom that such correspondences could be the product of coincidence—but many coincidences appear to be impossible, too." Now that he is a trained psychologist with a hard-earned doctorate, Harary has reached the conclusion that in the face of something "so inexplicable and potentially powerful, almost all of us, even scientists, can sometimes make the wrong assumptions or get confused." That cuts both ways, of course. But the saga of Dr. Keith Harary's journey through the psychic underworld might leave us dissatisfied, puzzled, and more than a little irritated. If even public advocates of parapsychology, formerly hailed as psychic superstars, can back out and put it all down to coincidence, why should the rest of us take the topic seriously?

Well, that's certainly the impression the CIA tried to convey when it pulled the plug on the Star Gate program two years after Harary's *Omni* interview. I hope I'm not considered unduly cynical if I admit that I don't necessarily accept this verdict just because it comes to us from such an authoritative and truthful source as the CIA. Suppose the word got out early in 1944 that the American military was working on nuclear weapons, and eager journalists put the question to a government spokesman. Naturally the candid reply would have been: "Yes, it's interesting that you ask that question—for some years now, we've been running a program called the Manhattan District project, which has met with quite some success. Miss Jones has some briefing papers here for you all, and a little later we'll have Dr. Oppenheimer in to talk with you." Or perhaps not.

The history of the U.S. psi research and applications program is murky and muddled, possibly by design. Or maybe just because of greed and opportunism. The most enjoyable journalistic book on the topic is still *Remote Viewers*, a 1996 paperback original by Jim Schnabel, an excellent, droll, and apparently well-heeled writer whose style falls between *Rolling Stone* and the *New Yorker*. His two earlier books about the fringes of the paranormal are equally witty and stinging, although they stand back from their subject matter (to the extent that's possible, given his own mischievous involvement in faking crop circles). His *Round in Circles* (1994) was a very funny study of these so-called crop circles, the enterprising characters who created them in the dark of the night, and those odd folk who are convinced that these ever more elaborate structures carved into fields of cereal must be the work of ancient earth forces or extraterrestrial visitors. *Dark White* (1995) looked more closely at those alleged extraterrestrials and their unpleasant habit of abducting and anally probing hapless humans; it is a hilarious, definitive study of UFO "abductions."

Given this sardonic background, one might have expected *Remote Viewers* to poke gentle fun at the lunatics who thought they could wander through space and time while lying on a couch, not to mention the lunatics in braid and uniform who paid them to do so. Not so. Schnabel had his fun, but found himself, to his surprise, rather taken by the evidence for remote viewing, and by the honest demeanor of most of the people he interviewed. Schnabel lists numerous eye-openers, some familiar from Targ and Puthoff's *Mind-Reach* (1977), others more recent. What must CIA analysts have made of the detailed drawings Pat Price provided SRI of a secret Soviet research and development target south of Semipalatinsk? Here in his sketches were the unusual, rail-tracked, gigantic eight-wheeled crane, the metal spheres being welded together, confirmed later by spy satellites. By the end of the book, which parallels a television program Schnabel made at the same time, *The Real X-Files,* he himself was learning the process and the protocols of remote viewing, with Ingo Swann at the helm. And surely to his faint horror—as if he'd discovered a flying saucer hovering beside one of the crop circles he helped to make himself—Schnabel produced the goods. He had become a successful remote viewer.

Not everyone is pleased with Schnabel's book. Some of the main players—scientists as well as the remote viewers themselves—find it patchy and occasionally absurd. Still, as an introduction to the SRI/military program, it remains invaluable, tracing the often convoluted history from Targ and Puthoff's Menlo Park initiative and Ingo Swann's development of the working protocols, through a blizzard of acronyms and code-names—INSCOM (the U.S. Army's Intelligence and Security Command), SCANATE (Swann's scanning by coordinate), Gondola Wish, Grill Flame, Star Gate—military officers such as General Albert Stubblebine (known to some irreverent colleagues as Stumblebine), who hoped to learn how to walk through walls, unnamed intelligence operatives, humble and not-so-humble soldiers invited to form the first squads of psychic espionage agents, the final collapse of morale and vision into a swampy mess of crystal ball gazers, card readers, astrologers, and superiors stuck with the job without much of a clue.

Undoubtedly the most reputable and admired of the military remote viewers is Joseph McMoneagle, who retired from the service in

1984 but was then hired by SRI as a research psychic. He's published several how-to books on remote viewing, two revealing memoirs— *Mind Trek* in 1993, and the more explicit *The Stargate Chronicles* (2002)—and a most unusual gathering of precognitive forecasts from the late 1980s through to the year 3000 and beyond, *The Ultimate Time Machine* (1998). The almost inevitable confusion of fact and hype in this publishing genre is evident in the cover art for *The Stargate Chronicles*, which shows at the top of the frame the nose cone of a missile presumably in its launch silo, with lamps, meters, and so forth surrounding it; in the lower half of the frame exactly the same detailed, complex image is being sketched by a hand holding a pen.

As a metaphor, this is cute but deeply misleading. As an accurate portrayal of the usual process of remote viewing, it is preposterous. Even the most remarkably accurate transcripts of expert remote viewers' perceptions are sketchy, partial, and filled with erroneous detail, just as likely to be inverted or skewed or indeed themselves metaphorical. (At any rate, those examples I've had the opportunity to read or to read about have been—although in a 2002 interview McMoneagle said that some still-classified work *is* this detailed.) This does not mean that they are useless. It does mean that long training is required to interpret the material, and in many cases the viewer does not himself or herself know what the target is, even after providing elaborate information that is unmistakable to those who know what they're looking for. In short, the cover of *The Stargate Chronicles* is a fudge, a blatant overstatement of what remote viewing can do. Anyway, what's with this word *Stargate* in the title? Isn't it *Star Gate*, two words? Yes, indeed, if we are to judge from McMoneagle's usage throughout the book, and from other informed sources such as Puthoff. Maybe there's a kind of popular fiction leakage from the science fiction TV show *Stargate SG-1*, itself derived from the space adventure movie *Stargate* (1994), which presumably borrowed the term from Sir Arthur C. Clarke's *2001: A Space Odyssey* (1968) and its sequels.

Meanwhile, other graduates from Star Gate, and their students after them, almost immediately began turning the already head-spinning process of remote viewing into a complete circus. Retired major Ed Dames set himself up, along with several other colleagues recently retired from the program (Mel Riley, formerly a photo analyst; Lyn Buchanan, a computer whiz; and others) as PSI-TECH, which started selling courses in remote viewing for many thousands of dollars apiece. Dames became notorious for his frequent appearances on the radio program "Coast to Coast," where he used his psychic powers to forecast one terrible disaster after another, none of them eventuating, although he managed to miss historic tragedies such as the destruction of the World Trade Center Twin Towers in September 2001. On the other hand, it's arguable that he helped to create another small tragedy: the death by collective suicide of a group of forty new-age UFO cultists, some of them self-castrated, known as Heaven's Gate, who slew themselves in the hope of being translated into a gigantic spacecraft that Dames and his associates remote-viewed as plunging toward the earth in conjunction with the comet Hale-Bopp in 1997. This was mostly the responsibility of a tenured professor of political science at Emory University, Dr. Courtney Brown, who was trained by Dames. This absurd tragedy is treated briefly in British commentator Jon Ronson's blackly funny *The Men Who Stare at Goats* (2005), where Brown rather vaguely disavows much memory of his participation. Meanwhile, after having a nervous breakdown where he spoke to angels, "Dr." David Morehouse, a Mormon major not quite honorably discharged from the army for a series of offenses that don't strike me as especially offensive—"sodomy" was one—published a book titled *Psychic Warrior* (1997) that Ed Dames claims is largely based on his life rather than Morehouse's.

It seems to me quite likely that the risks of plunging every day into the unconscious without a tether are considerable, and help explain the lunatic utterances of some of those formerly associated with these programs. Revealingly, Soviet psi researchers at a 1973 international conference in Prague, Czechoslovakia, asked Russell Targ, "What do you do to keep your subject from cracking up or going crazy during the experiments?" *(Mind Race,* 71–72). You can get the sense that delving into this weird unconscious material does put people at risk of, at worst, delusional psychosis. Dr. Edwin May commented to me: "I believe RV can be psychologically dangerous. Joe McMoneagle's uniqueness is mostly his capacity to handle his ability. I have had to dismiss otherwise sane people from the project because they 'went off the deep end' after doing too many RV's. I have no reason to doubt that this was also a problem for the unit at Fort Meade." May added with admirable candor: "I have a hard rule that I act as a subject once in each new experiment. I want to experience the protocol to see if it *feels* right. I think that in general, and for sure in my own case, it is too easy to become overwhelmed by ego after successful trials. I do well and it scares me. So I don't do it in general" (personal communication, July 5, 2006).

Schnabel's account of this history seems to me essentially believable (but see below). Perhaps it's because he stands carefully back from the fray, despite having learned successfully how to remote-view from both Swann and Ed Dames. He allows the deluded, hallucinated, crackpated, prankish, and felonious to reveal themselves, while playing fair with genuine intimations of mystery. If Morehouse were to be believed, by contrast, it is a matter of plain (if perhaps still-classified) record that Defense Intelligence Agency personnel performed psychic overhead flights of the first Gulf War, spoke to the dead, surveyed UFOs buzzing crop circles and under the sands of Mars, and locked on to the hidden Ark of the Covenant. Just as well that the major was out-of-body in that last-mentioned sacred place, for otherwise he'd have perished "instantly . . . vaporized" (Morehouse, 132). Desert Storm's oil-well fires, it turns out, were a cover-up for U.S. chemical weapons tests on the country's own soldiers.

Morehouse, out-of-body, saw the open canisters (169). That's why the army hounded him into court (and then let him go, after he ended up in the psychiatric ward of Walter Reed Army Medical Center, with an "other than honorable" discharge). Earlier, they'd tried to kill him and his family. Or maybe not. As even Morehouse admits, his book is heavily fictionalized, appropriating the experiences of others (some of them given proper or at least different attribution in Schnable). It was at one time said to be under lavish contract for the movies, perhaps to star Sylvester Stallone. Schnabel's *Remote Viewers* is at least an effective counter to Morehouse's vulgar concoctions.

So did the U.S. military program really develop effective methods for getting "into the ether" (lovely, dumb nineteenth-century term!), whizzing back and forth in space and time, contacting those toiling in the afterlife, angels, extraterrestrials in UFOs and on distant worlds? If even the least of Morehouse's assertions were correct—and Schnabel corroborated the least lurid of them—if his unit's methodology for psychic espionage actually did work, bolstered by data from regular spies, it's difficult to see why university parapsychologists at Edinburgh and Princeton and elsewhere kept fiddling about with their penny ante lab experiments or, a decade later, stood on the verge of closure.

Schnabel's 1995 British Channel 4 special, *The Real X-Files: America's Psychic Spies*, in gestures of postmodern mockery, lightly mapped his book's territory in a bizarre *Twin-Peaks*-meets-*American-Gothic* style. And while *Remote Viewers* was cautious and very fleeting on the topic of Morehouse's alleged exploits, Schnabel gave free vent to his distaste in an article posted widely on the Internet but now rather difficult to find:

> To me, his story is not just about the depths to which one human can sink. (Morehouse is, in the end, perhaps only a sleazier, crazier version of the old Sgt. Bilko character.)

Somehow his story also reflects the current state of things in
America, a country that seems to be going insane . . . What
else to call a people who feed hungrily, via *The X-Files* and
other forms of that hugely popular genre, on paranoid con-
spiracy fantasies otherwise found only on psychiatric wards?

Morehouse's own account of his life shows an honorable man
tragically misused by evil manipulators, but saved from suicidal
depression by eternal love for his angelic wife plus some actual
kitschy angels and his patriotism and decency. It's intriguing to compare this with Schnabel's relentless savaging in his article, which looks
rather like a chapter excluded mysteriously from his book: Morehouse the scummy womanizer, the guy with "a small home improvement business, House Tech, that he ran on the side." Can this be the
same Morehouse who rails against the shocking state of morale and
morals in the spook business (the other sort of spooks)? "And some
of their private lives! A few staff members hardly bothered to conceal
extramarital affairs. . . . It was a real Peyton Place, and I hated it. . . .
Let me top my description off with the colonel who sold Afghani rugs
out of the trunk of his car in the parking lot" (Morehouse, 157).

The major's alleged sexual improprieties, even his goldbricking
while on the payroll of the Defense Intelligence Agency (DIA), are
finally irrelevant. We all have our little quirks. It is notable, though,
that none of this material was touched in his own book, not even to
rebut it. The jacket mentions his PhD (in education administration).
"My doctorate is from LeSalle University in Mandeville, LA," he told
me by e-mail. He meant La Salle. In a rather odd disclaimer, he
added: "I did not simply buy my degree—believe it or not, I had to
work for it" (personal e-mail communication). It would be easiest,
obviously, to discount Morehouse as a former, or feigned, psychotic
with schmaltzy style and delusions of grandeur. But we are meant to
accept that this kind of operative provided the psychic material that

the CIA and DIA and other dark powers once found impressive. Something about all of this stinks to me of a disinformation package, rich in odors attractive to new age nostrils and old *X-Files* fans, but just as certain to wrinkle the snooty noses of skeptics.

Meanwhile, former military remote viewers such as Lyn Buchanan, Ed Dames, and Joe McMoneagle, some still selling ambitious coaching and psi-for-hire programs, are represented on the Internet. Dames, having been expelled from PSI-TECH by his former wife, runs the Hawaiian company TRV Institute, which advertises:

> There are actually only a handful of Remote Viewing instructors in the world that were officially training with the military developed RV protocol [sic]. If you find an instructor not on the list below, stay away!
>
> **Major Ed Dames**
> **Paul Smith**
> **Lynn Buchanon** [sic]
> **David Morehouse**
> **Psi-Tech Inc.**
> **Aaron Donahue**

Buchanan's site, P>S>I> (Problems> Solutions> Innovations>), claimed that perhaps because of the benign influence of morphic resonance, remote viewing skills can currently be learned with astonishing rapidity. In just a few concentrated days or weeks, for a hefty fee, students can master powers over time and space that used to take months under Swann's structured protocol (*CRV: Recent Developments in Remote Viewing training,* on the P>S>I> Web site). Swann's way was a sequence of neologized stages:

One, minimal sketched "ideograms";

Two, basic sense perceptions and emotional impacts, progressing through "dimensionals," to:

Three, large autonomous sketches;

Four, listing of "tangible" and "intangible" quantities;

Five, prompting for detail from specific target objects;

Six, building a solid model in clay, like the UFO-obsessed seers in *Close Encounters of the Third Kind* or perhaps, to be fair, like an architect externalizing her blueprint;

Seven, uttering free-floating phonemes in order to elicit meaningful cues and clues; and so on. (See Schnabel, 242–52)

By his own account—perhaps one might ponder the degree of its exact truth—Swann is a prodigy who had read the entire thirty-volume *Encyclopaedia Britannica* before he entered kindergarten, and flow-charted the *Tao-te Ching* when he was seven (*Real Story*). No less polymathic was the scientific mind of Hal Puthoff, who, you'll recall, in 1985 would walk away from SRI to research a fabulous alternative paradigm in deep physics, "zero-point" or "vacuum" energy, a quantum alternative to general relativity. One has to wonder, did Puthoff use his disintegrating RV team's somewhat unreliable powers of precognition to select this risky venture? Puthoff does not say, nor do any of his former associates to whom I've put that question, but military remote viewer McMoneagle has written, "My gut tells me that he is probably on to something, and certainly remote viewing helped" (1998, 111). It is intriguing that Puthoff did employ "associational remote viewing," coding future market movements to binary targets, and reportedly earned twenty-five thousand dollars in a month for a community school (322). Any lingering doubt as to the true efficacy of the RV method might fix itself to this kind of success story. If Puthoff and his team of charitable RVers could precognize the stock market, and if, in a famous earlier debacle, silver futures had been traded with some success (Schnabel 265, 322), why, then, are we still arguing over the merits of the method?

If government drug busters, armed forces, and CIA, DIA, NSA, and who-knows-what covert agencies attained some measure of astonishing accuracy in RV tasks—as Schnabel, Morehouse, McMoneagle, Swann, and Buchanan assure us they did—why were these programs allowed to whimper away? Why were they closed down as their early champions fell from grace or retired—men so well placed as General Bert Stubblebine, chief of the army's Intelligence and Security Command; General Ed Thompson, the army's assistant chief of staff for intelligence; and Dr. Jack Verona, head of the DIA's Scientific and Technical Intelligence Directorate? Is it enough to blame the resistance of stolid scientific paradigms, the contamination of lunatic cultisms, or the ebb and flow of social currents like the complex blend that saw nuclear power lauded one decade and despised the next?

In any event, at least one former member of the remote viewing unit, an INSCOM officer, has vouched for Schnabel's account: "It appears to be very accurate. . . . While I do think Joe McMoneagle wrote a good RV book [*Mind Trek*], he was restricted in what he could say at that time. Schnable has certainly done his homework." But can such allegedly classified details truly be reliable? I asked Lyn Buchanan, and he kindly provided this comment:

Jim Schnabel's book is probably the only one available to the public which sticks closely to the truth and real history of the unit. There is a lot which is totally in error—in spite of the fact that it was information he should have had an easy time uncovering, and a lot which is surprisingly correct, in spite of being information he shouldn't have been able to get. It seems that he got many of the easy facts wrong, and really aced the stuff that should have been highly classified. At any rate, it still sticks most closely to the truth and real history. In summary, it isn't a totally accurate account, but it's the best available at the moment. Anyone wanting to know about the unit would be best to start here. (personal communication, 4 November, 1997)

Edwin May was more forthright:

> Schnabel's book is mostly a work of fiction. His error is that, while he bragged about his independent sources for interviews, what he never realized was that they were nothing at all like independent. Further more to write a "definitive" story about the project and not request an interview with the project director for the last 10 of 20 years [that is, May himself] seems a little odd at best. He slandered (indirectly) Jack Verona so much so that Jack got permission (thus I speak of him here) to write a blistering letter to the newspaper that first published excerpts. Schnabel reported that he was the recipient of a psychic blow job! Indeed! Terrible book. (Personal communication, 11 June, 2006)

Clearly much the same can be said of Morehouse's soufflé. In any event, to me, as an open-minded outsider agog at the reported accuracy of the best cases cited by RVers, the obvious gnawing question remains: If this methodology can optimally work perhaps 50 percent of the time, especially when bolstered by signal-enhancing measures such as *redundancy* (tasking several viewers without their knowledge at the same target), how could such power be left to slip away?

"Even if a remote viewing of a specific target proved to have been accurate and useful," Schnabel concluded, "its user, afraid of the giggle factor, could almost never acknowledge the fact" (338). Prejudice and paradigms, the usual suspects. I can't buy it. If the famous medium D. D. Home returned and flew through the air in front of video cameras, wired to the eyeballs with slick neuroscience electronics, I suspect he'd be whisked away promptly to a military lab, giggle factor or not. If Mel Riley and the rest really did pluck data from the vasty deep, how could any paradigm resist? I disbelieve in iridology and homeopathy in part because my instinct tells me that if these laughable ideas endured the sieve of scientific testing, by now they would be the proprietary lines of Swiss billionaires. Same with remote viewing. I'm not so much dismayed by the coarse, hallucinated mythos its shamanic practitioners seem to generate as they "bounce off the walls" (Schnabel, 73ff)—the pregnant Martian females under the New Mexico desert (377), the visions of the Ark of the Covenant—as I am flummoxed by their failure to have bought out Bill Gates by now.

But perhaps there are reasons for this information—given that it has escaped from custody—to be available mostly in popular books, Internet scams, and comfortingly scary movies and TV series. The way in which the psi programs were wound up is extremely revealing, when you stop to think about it for a minute or two.

෴

Drs. Jessica Utts and Ray Hyman, both of whom had previously assessed psi for the military in 1988,[3] were contracted in 1995 by the American Institutes for Research (AIR) to evaluate the Star Gate program. Under Pentagon pressure, the multi-decade program was then terminated late in 1995; one might suppose that these evaluators had examined the best results from the various wings of the lengthy project and found them wanting. Not so, or not quite. Curiously, Utts's report had concluded: *"It is clear to this author that anomalous cognition is possible and has been demonstrated.* This conclusion is not based on belief, but rather on commonly accepted scientific criteria" (my italics).

Hyman, a well-known skeptic, predictably disagreed, but admitted: "I tend to agree with Professor Utts that real effects are occurring in these experiments." Irritatingly, the assessors were not given the "operational" remote viewing material, just laboratory results from the last three years of government-sponsored research done from 1989 to 1994 at Science Applications International Corporation (SAIC), after the program had moved from SRI. Most of the real material, according to Utts, remains classified (personal communication), although Ed May is doing what he can to release it. We'll come back to this disturbing piece of media manipulation.

The subtitle of Jim Schnabel's account dubbed it "The Secret History of America's Psychic Spies." Morehouse's very strange book was subtitled "The True Story of a Soldier's Espionage and Awakening." Ingo Swann, arguably the eminence grise of the defunct official RV programs, for years has been releasing chapters of his own history on the World Wide Web (after it was declined by America's "top five mainstream publishers" due, incredibly, to a perceived lack of public interest in the paranormal). It provides a quirky if unfinished triangulation on the issue. "If nothing else," Swann wrote, "the American intelligence community is VERY mainstream. The odds of a 'psychic' (as I unfortunately was to be dubbed) of even entering into the realms of mainstream structured international espionage, much less surviving for some eighteen YEARS within the abundant machinations—well, such odds were nonexistent as of 1971" (chapter three, his uppercase). Weirdly written in a voice that yokes *La Cage Aux Folles* to stilted bureaucratese—the Vietnam War, he remarks, led to "the widespread realization that the rationale for that war was nutso whacko"—this jumbled blend of immodest autobiography and strange history is unabashedly titled *Remote Viewing—The Real Story*.

How "real" and "true" we should deem versions of that "secret" story is perhaps up to each reader, and opinions will vary considerably. Meanwhile, the methodology and protocols of remote viewing are now very much a matter of public record. Why hasn't anyone else used them to remarkable effect?

Well, as it happens, someone has. Let me tell you about the detailed psychic discovery of Iraqi despot Saddam Hussein, on November 3, 2003, six weeks before "Operation Red Dawn" troops pulled him, disheveled, bearded, and dispirited, from a hole in the ground near his hometown of Tikrit.

3.

WATCHING SADDAM FROM HOME

For *Naked Science*, telepathy gets the thumbs down. Even if we accepted it, no one can tell us how it works, or why. Until there is incontrovertible evidence, it must remain on the fringes. Fascinating? Yes. But a real science? No!

—*Naked Science: Telepathy*, National Geographic television

Placebo drugs work despite the fact that they have no active ingredient— or rather, the active ingredient is the patient's own immune system, spruced up by an act of belief. If psi exists as a genuine human potential, there might be an odd consequence: psychic snake-oil salesmen and dubious teachers of arcane disciplines could prove surprisingly effective. An analogy that comes to me, because I am a writer by trade, are those expensive courses offered by hopeless hacks who promise to teach you how to make big money as a novelist. Now it's true that some new writers do indeed make big money, even if they have to wait many years until their skill has developed, and their reputation alongside it. But expensive tuition, even when it's given by experienced and capable writers, can't turn a boring, tone-deaf plodder into Vladimir Nabokov or even Dan Brown. Well, maybe Dan Brown.

But suppose some flashy salesman's goods amount to nothing more than a set of basic instructions: Read a lot of books. Buy a pad and pen, or a computer. Let your mind roam imaginatively into the realm of story. Open the pad or computer screen, and start writing down your imaginary scene. Keep doing this every day, stopping now and then to read what you have written in order to assess whether it has any resemblance at all to the kind of published work you admire. Press on until you have produced at least a million words. If you diligently write five hundred or a thousand words a day, this will take you somewhere between three and six years. At the end of this ordeal, assuming you manage to sustain the effort and your conviction in your own talent, there is a reasonable chance that you will be ready to produce a worthwhile book. That's all. The other pamphlets, handbooks, guides, dictionaries, thesauruses, and use of Google to check facts, and lofty advice from the professors . . . all are ancillary. You could be Robinson Crusoe on his island, half naked, with just a notebook and a box of pencils, and this banal advice would have a good chance of carrying you through to modest success.

It wouldn't matter, you see, whether this fairly rudimentary counsel were served up by a Nobel Literature Prize winner or a scam artist down on her luck, hoping to snare some dollars from the gullible. The point is, we humans *do* have a capacity to imagine, most of us, and a potentially unlimited power of utterance. Whether any particular person produces something more than routine chatter or gossip is, to a large extent, a blend of choice and innate talent. Most writers teach *themselves* how to write, learning by reading and doing, with a little help from those of their friends who are willing to read their effusions and tell them, as kindly as they can, what works and what stinks.

Here's where my analogy cashes out: if psi is real, if the human mind truly has the capacity to reach beyond the accepted limits of our senses—even when those senses are augmented by technology—it follows that remote viewing probably *can* be done by many of us, with at least some degree of skill, whether or not we are trained in its use by experts or by scoundrels.

On the other hand, we also know that to master Olympic-level gymnastics, printmaking, or the violin, it helps—indeed, it's probably imperative—to have the guidance of an expert as well as the supportive or bracingly contestatory fellowship of other beginners. There's a hazard, however, if the skill that's learned is evaluated by purely subjective standards. At the extreme, you run the risk of what we might call *delusional reinforcement.*

Mutual criticism in writing groups can end up in a pleasant haze of equally mutual congratulation, never putting the work to the brutal test of submission to a commercial market, or of finding excuses if it fails there. In the case of remote viewing, or other would-be psi applications, the equivalent might be the sort of horrifying cascade of delusional nonsense emerging from small groups "trained" by Ed Dames and Courtney Brown, persuading one another that their roaming minds had locked on to spacecraft of redemption or terrifying near-future, solar-flare planetary disasters. And yet these gruesome misfirings might be nothing more than fabrications of people placing themselves into mutually reinforcing fantasy states. If psi is real, it seems to me very likely that such pathologies can be an uncontrolled side effect of genuine experiences gained at the start of the learning curve, when people attempting RV actually do find themselves, to their amazement, locating lost keys in unexpected places, drawing accurate scenes witnessed by "outbound beacon" friends, capturing in advance the next day's cartoons in the *New York Times* or the local newspaper.

Here's an absolutely trivial example, meant only for illustration and certainly not as scientific evidence of anything. Today I walked in the hot, early afternoon Texas sun to my wife's office, to collect a UPS-delivered package. I was expecting a computer disk with data on it. This package, though, was fairly obviously a book. Who had sent it to me? On quick inspection of the padded envelope, I could see no indication. (I'm significantly sight-impaired, needing quite different prescription glasses for reading and long distance; this time I was wearing the long-distance specs, which blur out small print details.) I review science books for a national newspaper, and I had recently requested the following titles from my editor:

Susskind, Leonard, *The Cosmic Landscape: String Theory and the Illusion of Intelligent Design* (Little, Brown)

Woit, Peter, *Not Even Wrong: The Failure of String Theory and the Continuing Challenge to Unify the Laws of Physics* (Jonathan Cape)

Smolin, Lee, *The Trouble with Physics: The Rise of String Theory, the Fall of a Science, and What Comes Next* (Houghton Mifflin)

Might it be one of these?

As I sauntered home, sweating in the hot sunshine, I opened myself up to the possibility that if psi exists, I might be able to use it to detect something about the book still sealed tightly and stapled inside its heavy brown envelope, perhaps something of its appearance. I don't do this very often, but on those rare occasions when I do, I'm quite often right. What came to me was the idea—not a visual image as such—that it was a red book, with horizontal stripes across it (I had in mind a series of racy lines perhaps an eighth of an inch wide). Title? Something about God. That was fairly clear. Hmm. All I remembered of the titles that I'd sent to my editor was that they had dealt with high-energy physics, a topic that has been known to attract the word "God" into the title for absurd but highly saleable motives.

There was also the faint possibility that it might be a recent book by philosopher Daniel Dennett or evolutionary biologist Richard Dawkins on the sources of religious belief. I was fairly sure that the Dennett book had been reviewed already by somebody else.

I got home, played with the dog for a minute or two, slurped down several glasses of cold water, and finally ripped the package open with some difficulty. The cover was two shades of red, with a lighter shade in a strip across the middle. The title was shown thus:

the
improbability
of

G O D

edited by
Michael Martin
&
Ricki Monnier

The publisher was Prometheus Press. (When I looked more carefully at the envelope, wearing my reading lenses, I found the sender's name and address in gray, tiny four- or five-point print.) I couldn't recall hearing of this book, but I knew I'd probably requested it from Prometheus months earlier. Let me now go into my e-mail records. Why yes, indeed, back at the end of March 2006, I did submit that request (along with two quite different titles, one of them concerning flowers). So it's possible that some diligent memory module of my brain had accessed a list of all the books I've asked for in the last year or so, and also recognized with unconscious acuity, from the package, that it came from one particular publisher. Here, then, is a skeptical explanation: I unconsciously recognized that publisher, Prometheus, from the parcel, so the probability of this particular book having God in the title is 1 in 3, since I'd asked them for three books. Covers tend to be predominantly white, black, red, blue, green, or yellow, or of course they might use a distinctive image containing many different hues. But let's say 1 chance in 6 (the number of colors), conservatively. As for the stripes—well, they weren't the kind of stripes I had in mind, but at least they were horizontal rather than vertical or at a slant, so let's say 1 in 3 (the number of possible positions). What, then, is the likelihood of my getting this much right purely by subconsciously informed chance? On this exceptionally crude probability estimate, it's 1 in 3 by 6 by 3, or 1 in 54—better than 0.05, but not as good as 0.01. This is perhaps slight evidence for psi, then, but not very impressive. On the other hand, I can assure you that it *felt* rather striking, and confirmatory.

In other words, nothing elaborate is needed in the exercise of remote viewing, at least at this extremely rudimentary level. That much was recognized back in 1930, in the Upton Sinclair book *Mental Radio,* and more explicitly in 1948, when French writer René Warcollier published his book (in French, of course) *Mind to Mind.* For Warcollier, the secret of cognitive psi was that:

> There seems to be an analogy between the paranormal perception of drawings and the normal perception of drawings exposed for very short periods of time by means of a tachistoscope [a machine that flashes up images very briefly]. . . . There is very little doubt that the use of language can cause difficulty in receiving a telepathic impression, because the medium of expression in telepathy is not language. If what is perceived seems illogical and irrational, the chances are greater that it is a hit than when the recipient restricts himself to certain persons, places, and things.

If all this is so, we can see why almost any systematic approach to perceptual psi might work, even if it's sold for a preposterous amount of money to the gullible. In fact, placebos often seem to work best if you have to pay the doctor a steep fee. It's conceivable that you would gain an equal benefit from someone pretending to have served in an important role in the Star Gate program, teaching some pompously rebranded variation of the methodology devised by Ingo Swann or others. On the other hand, why go to a cult or a con artist for illumination? Fortunately, not all of those who promise to teach the methods of what we might call *applied psychic technology* are crackpots or scam artists, although in my experience some of them do have rather offbeat philosophic and religious opinions. Just as the psychical researchers at the start of the twentieth century tended to fall under the sway of spiritualism, I have noticed that many contemporary psi researchers are metaphysical dualists who refused to countenance the reigning "reductionist" monism of the scientific community—that is, they insist that the only way to explain psi is to acknowledge an order of existence separate from the material, a "spiritual" order of reality, perhaps more fundamental and certainly more meaningful than the violent and cruel if often beautiful world of space, time, and causality. Others are metaphysical monists who take the large step of claiming, as idealist philosopher Bishop Berkeley famously did, and as Buddhism does, that the material world is a kind of illusion, or *maya,* and that reality, beneath this deceitful surface, is a realm of Will and Idea, or something even more ethereal and mysterious than that.

My friend Stephan A. Schwartz, a former research director for the Rhine Research Center, seems to fall into this latter class, to my vexation, but his success in remote viewing is remarkable. His deep conviction is that everything is nonlocally connected, and that this is why we can extend our attention and intentions beyond the reach of our senses and our limbs, touch the unknown in quite practical ways that fetch back to us information about what is lost to memory or has not yet happened, and bring healing to the ill. He comments:

I believe the reason my research has tended to work is that I have tried to conform to what I believe I know about how nonlocal consciousness works, based on my research, and the research of others for the past half century. I am a meditator, so I have some direct experience with this state of consciousness, and I have always seen an experiment as an *intention contract* in which everyone who shares the intention to participate is a player in the process. Based on my studies of spiritual traditions, which I see as the precursors of parapsychological research, I believe the traditional model of experimenters and subjects is a pernicious fiction. Blindness and randomization are important for reasons other than those normally assumed. Nothing is blind, but blindness helps with metal clutter, the enemy of good nonlocal awareness. (Personal communication, October 23, 2006)

As it happens, I had never met Stephan until recently, but fortunately, this is the twenty-first century, and even though telepathy is not yet routine, there is always e-mail, instant messaging, and the blessings of search engines that can scour the World Wide Web in seconds. It has been my good fortune to communicate with many of the most notable individuals studying and investigating psi anomalies—some call themselves parapsychologists, others prefer not to—via private e-mail lists where the rule is rather like the first and second rules in the movie *Fight Club* ("Do not talk about Fight Club"). Exchanges on the list are confidential: do not quote verbatim, do not attribute. This restriction is very frustrating for someone writing about psi (or about any interesting and controversial topic, for that matter), but of course I absolutely understand and abide by the necessity. No confidences have been harmed in the making of this book. Specialists need to be able to let their hair down in private and speak to one another with a degree of candor that might not be permissible in the faculty staff room, let alone in the press. So all I can do here is mention my gratitude at being permitted to audit these conversations, and sometimes to contribute to them. It has given me an unusual and privileged glimpse into a complex, sometimes ornery backroom where psi researchers speak candidly to one another about their latest findings, their disagreements, and occasionally their philosophies. I met Stephan Schwartz on one such list, and found his adult and responsible voice especially refreshing in a context that is so often pilloried by the skeptical as a happy hunting ground of fakes and the deluded.

Stephan is one of the long-term explorers of remote viewing, having initially developed his approach independently of the military and intelligence programs, although subsequently in fruitful cooperation with their principals. It was Schwartz whose contacts in the U.S. Navy led to his borrowing for three days the submersible *Taurus*—a small submarine—from a Canadian deep-ocean research corporation, Hydrodynamics Company, who were doing sea trials off the Los Angeles coast at the institute's research facility on Santa Catalina Island. Using that submersible, Project Deep Quest carried Ingo Swann and Hella Hammid 558 feet below the surface, radio switched off, well out to sea on a rough July 16, 1977, to learn if they could remote-view targets back on land. (Hal Puthoff and Russell Targ were also involved, on land.) This was a critical experiment, because success in these tests would mean that psi could not be explained as a form of electromagnetic radiation. Submerged under the sea, they were shielded against all known kinds of electromagnetic radiation capable of transmitting messages to a human brain. Nevertheless, although poor Hella was on the verge of nauseated seasickness, both she and Ingo Swann achieved their targets—which allowed the experimenter to identify a particular code prepared in advance, manifesting the kind of magic communication that the commanders of nuclear submarines might need if a terminal emergency arose and they needed to communicate with headquarters without coming to the surface.

Here was Hella's verbal response: "A very tall looming object. A very, very huge tall tree . . . a cliff behind them. . . . Hal is playing in the tree. Not very scientific." A series of six possible targets, each encoding or representing a standard message, had been carried in a sealed envelope on the submersible, twin to the envelope of optional targets used by the outbound beacons. Her responses complete, and with no feedback from the shore, Heller examined the six possible images, and immediately selected a large tree on the edge of the cliff. Subsequently this proved to be the site to which the outbound beacons had traveled. What is impressive about this is not just that Hella had selected one correct option out of six, but that the free-form detail was *so* remarkably accurate. The competing images could have been drawn, after all, from almost any place in Los Angeles.

Sometime later, in his turn, Swann was taken down 256 feet. He reported that Hal and Russell were now walking about in a large enclosed space, possibly city hall, no, a mall, with reddish, flat stone flooring. The closest possible target in the fresh array of six, sealed in the second envelope, was the Old Mill shopping mall in Mountain View, California. This was correct also, and both guesses were later blind-matched successfully in first place by an independent scientist.

What interests me especially about Stephan Schwartz is the range of his interests and the breadth of his experiences; he reminds me rather of that nineteenth-century literary adventurer Sir Richard Burton. After early involvement in the civil rights movement, as a young civilian in Washington, D.C., Schwartz wrote speeches for President Richard Nixon on such matters as the law of the sea (an international treaty negotiated while Nixon was in office), the development of an all-volunteer military, and the future role of the navy. In a 2005 interview he commented: "In the '70s, I was part of the small team that transformed the American military from an elitist conscription organization to an all voluntary meritocracy, the military we have today, which doesn't care what race you are, or where you went to school, or who your family is, or how much money you've got. Colin Powell is the epitome of this. He is a foreign born, non–West Point man of color, who rose to be the senior military officer in the United States and, then, Secretary of State." As the Iraq War and insurgency moved further into murk, and generals spoke out without much effect against the views of the Bush administration, it was not clear even to precognitive remote viewers whether this success would be a lasting one (Interview, Horrigan, 2005). He wrote articles and speeches for two chiefs of naval operations, Admirals Elmo and James Holloway; for James Schlesinger, secretary of defense; and John Warner, when he was secretary of the navy. As civilian special assistant for research and analysis to Admirals Zumwalt and Holloway, normally a post held by a ranking naval officer, he earned a certificate of commendation for his efforts. He told me: "I worked for several years, at the height of the Viet Nam War, a war I despised, and wrote hundreds of thousands of public words for various principals, yet never was asked to write a word about Viet Nam. Everyone knew my feelings, I guess, and just got someone else to do it, or winged it" (personal communication, August 30, 2006).

Stephan has been on the editorial staff of *National Geographic* and has published in the *Smithsonian*. In the 1980s and '90s he was involved with citizen diplomacy between the Soviet Union and the United States, helping create back channels of information between their governments, and involved as well in philanthropic ventures. He has led a number of audacious and successful expeditions in search of lost treasures such as Marc Antony's palace in Alexandria, the remains of one of the ancient wonders of the world, the Pharos Lighthouse, and numerous sunken vessels. He has written books describing these discoveries (*The Alexandria Project*, 1983, revised

2001) and the application and history of remote viewing techniques used in finding them (*The Secret Vaults of Time: Psychic Archaeology and the Quest for Man's Beginnings,* 1978).

But wait, there's more! Incredibly, with several friends he was the inventor of that amazing infomercial fad of the 1980s and '90s, the ThighMaster-Exerciser (just $14.99—call this 800 number *now*). The ThighMaster made $100 million for marketers. At that time Stephan Schwartz was running something called the Mobius Laboratory, devoted to exploring psi and consciousness, and also Clearlight Productions, a production company that made network programming. He and his friends sold their interest in the device, which has the disturbingly bent shape of a Möbius strip. TV promoters preferred to call it an infinity sign, or a butterfly's wings. Stephan told me: "The design was at once an inside joke, and the most efficient possible design for manufacture. It got cooked up one day over lunch amongst myself and three friends. Before I would let it be marketed we gave grants to the sports medicine department of the schools of medicine at UCLA and USC, and they independently ran studies, so I knew that all claims were supported by actual research" (personal communication, August 30, 2006). (An entertaining but somewhat *gappy* history of this infomercial megahit is merchandiser Peter Bieler's *"this business has legs,"* which ungratefully claims that "the bunch of people who created the ThighMaster couldn't engineer their way out of a paper bag . . . they were flakes . . . a television producer, a psychic advisor, the owner of a health spa, and an entrepreneur with a bachelor's degree in psychology. Not an engineer among them" (148–49).

Schwartz holds a particular interest in psi-mediated healing and has participated in experiments of "therapeutic intent" and similar nonconventional adjuncts to standard allopathic medicine and surgery. Tragically, this complementary approach failed to save the life of his wife, Hayden, who died several years ago. (Let me stress that this is not a sardonic insinuation that such methods are useless and self-deceiving;

sometimes they are, when used exclusively and to the detriment of conventional treatment, but the best hospital treatment fails us all in the end.) A longtime meditator and mystic of sorts, Stephan accepts this awful loss of his beloved wife as, I believe, a kind of passage into a nonlocal condition where they will meet again after his own death—or perhaps, to the extent that it is timeless, might be said to have already done so. Whether or not there is any validity in this ancient and widespread belief (I doubt it, although Stephan's Web site bibliography of refereed papers attests to healing via distant intent[1]), his success in eliciting remote viewing prowess in his students is quite astonishing. Which brings us finally to the strange case of the newbie remote viewers and the cowering dictator, Saddam Hussein.

On the ninth of April 2003, some three weeks before his sixty-sixth birthday, the brutal authoritarian president of Iraq was overthrown when the American-led invasion deposed his government. Saddam disappeared, and although his two sons were intercepted and died resisting capture, the father of the nation was nowhere to be found. Many assumed he had fled to Europe; he had been salting away hundreds of millions of dollars of Iraq's oil wealth. Hefty rewards were posted for his capture. We know now that Iraqi troops were deserting Saddam as rockets fell nearby and he arranged cars to take his two eldest daughters to Syria. He drove with two bodyguards in a white Oldsmobile to a bunker on the northern outskirts of Baghdad. For months he moved from one safe house to another by taxi, dressed as a simple shepherd. In July, when his sons were cornered and killed in Mosul, an instant legend arose that Saddam visited their graves that evening and wept for them. Over the following months, his traditional power base was entered by U.S. troops, his closest followers rounded up, and it seemed only a matter of time before the dictator himself was found. Even so, after six and then seven months, still he was nowhere to be found.

If psi remote viewing is so wonderful, although admittedly somewhat haphazard and unreliable, how was it that nobody inside or outside the military had tracked this evil fellow to his lair?

Stephan Schwartz, who lives in Virginia Beach, Virginia, was teaching a remote viewing seminar at nearby Atlantic University on November 3, 2003, and it occurred to some of his students, all RV novices, that tracking Saddam would be a very interesting and informative exercise. After a three-hour presentation of remote viewing techniques, Schwartz had invited his forty-seven novice participants to project their awareness elsewhere in space and time, to conceive of themselves standing full size but invisible and gazing at—what target? And where? The students themselves provided an emotional suggestion: the scene of Saddam Hussein's eventual capture by Allied forces, where he *would be* located at that future moment.

"Okay, you're standing there," Stephan told them, "you're with Saddam Hussein, you're life size—where are you?" They were instructed not to make judgments about what they saw; just try to capture the sense impressions they would experience if they were actually at that location, witnessing this pivotal moment in near-future history. For the next fifteen or twenty minutes they did just that, following the protocols in which they had been instructed, starting with a sketch of a small geometric image to "fix a contact with the information channel," then proceeding as if they were invisible eyewitnesses making notes, checking through their senses: taste, touch, smell, sight, the quality of light, the clothes the man is wearing, his appearance, and so forth. The hope was to reconstruct the event— even though it had not yet happened—as accurately as a court or a journalist might reconstruct a crime from the eyewitness testimony of many people catching glimpses from different angles.

Of course, it was possible that Saddam had long since escaped to Europe or Syria and was enjoying the immense fruits of his embezzlement of Iraq's oil wealth. Several participants did reach that conclusion, perceiving the dictator sitting comfortably in suit and tie beside a sparkling blue swimming pool. It was also possible that Saddam was dead, perhaps buried in a nameless grave by one of his last dutiful attendants. The key to remote viewing, Schwartz had told them, is to make no judgments, to banish analysis, to remain open to vagrant impressions. The students did just this, and at the end of the session their data was collected, photocopied, and the copies distributed, while the originals were sequestered by the administrator of Atlantic University, Herk Stokely, acting as an independent third party. In front of a notary, Stokely sealed the documents from that session into an envelope and locked it away in the vault of the Association for Research and Enlightenment (more on that group later), where it resides to this day, untouched and awaiting some final exercise of validation—an "unimpeachable chain of chronology."

That was it for the students. They had no immediate feedback, obviously, since nobody knew, in early November 2003, where Saddam was hiding, although some suspected that he was already in custody and that U.S. officials were awaiting a propitious time to bring him out in front of the reporters' cameras. Had the Star Gate program still been running, they surely would have been tasked to undertake precisely the same search. It is not impossible that some psi espionage "black program" does continue to operate deep undercover; if so, perhaps their endeavors in this regard must have produced results not unlike those in Virginia Beach.

Stephan Schwartz took the photocopied material and immediately analyzed it on pretty much the same basic principle employed by his psi application research since the 1970s—that is, he looked for consistencies, redundancies, and consensus among the multiple viewers. If only two people saw Saddam lolling by a swimming pool while the rest detected him three feet under the sandy soil, dead as a mackerel, then Schwartz would estimate that the latter projection was more plausible, carried more weight, than the former. Not that this is always a safe bet, according to the findings of the former SRI and SAIC researchers. Joe McMoneagle once said that if two people offered a particular remote view while five others agreed on a completely different scene, chances were about even that the two were right and the five were wrong. Psychic perception can create its own interference. This makes better intuitive sense if you compare the act of attempted psychic perception to a sort of "Chinese whispers" rumor, with everybody not quite hearing correctly what everybody else is saying, while muttering their own interpretations, so that perhaps a strong misinterpretation carries the day over a weak correct interpretation. (Think of the classic Sermon on the Mount scene in Monty Python's *Life of Brian,* where ardent but baffled disciples persuade one another that their master has offered the unlikely beatitude, "Blessed are the Cheesemakers.")

So not all that much can be done to ward off psychic cross talk or contamination, but Stephan had a couple of tricks up his sleeve. True, he was privileging the data points where information agreed, taking their consensus vote as a best-guess estimate of one feature or another. In addition, though, he sought "low *a priority* observations." These would be unexpected details that were hard to predict, unusual features of the residence, if there was one, anything striking or unusual in his appearance—say, if almost everyone agreed that Saddam would be discovered dressed as the fairy godmother in a Cinderella pantomime (*my* example, I should mention; Stephan was not being whimsical about this grave topic).

Here's what turned out to be the central consensus projection, based on those November 3 stray imaginings:

> Saddam Hussein will be found crouching in a subterranean
> room or cave, which is reached by a tunnel. It will be beneath
> an ordinary-looking house on the outskirts of a small village
> near Tikrit. The house will be part of a small compound that

is bordered on one side by a dirt road and, on the other, by a nearby river. There will be vegetation, including a large palm tree in the area.

Hussein himself will look like a homeless person, with dirty rough clothing, long ratty hair, and a substantial and equally ratty salt-and-pepper beard. He will have only two or three supporters with him at the time of his discovery. He will have a gun, and a quantity of money with him. He will be defiant but will not put up any resistance; in fact he will be tired and dispirited. (News, *Venture Inward,* Mar/Apr 2004, p. 30)

You have to ask, how much of this might be self-evidently likely at the start of November 2003, more than a month before a commando force tracked down Hussein, with the guidance of information presumably tortured (or bribed) out of one of his followers? I suppose it's a roll of the die among the possibilities. Saddam is rich and pampered, well beyond the borders of Iraq; Saddam is harried but spick-and-span in military uniform, leading the beginnings of an insurgency against invaders; Saddam is reduced to pitiful ruin, on the run. Perhaps the third option seems the most plausible, at least to patriotic Americans. Where, then, is he likely to be found cowering? Most commentators at the time seemed to think Saddam would retreat to the intense family loyalties of Tikrit, so that is not such a stretch. But what *is* remarkable about these predictions is their specificity, some of which only becomes apparent when you compare the eventual reality against the sketches of the house where the remote viewers expected him to be found.

As we now know, Saddam was indeed captured on December 16, looking precisely like a homeless man, despite the $750,000 in $100 bills he had with him and the pistol he carried on his person. Famously, he was found cowering in a "spider hole" beneath a fairly rudimentary house perhaps typical of the region but, to my eye, of unusual design (by Western standards, that is, which is why the accuracy of the drawings by ordinary American remote viewers is so startling), reached by a crawl space with an air vent that was included in a sketch by one of the remote viewers. He seemed dazed and offered no resistance. Some commentators, as noted, have speculated that he was drugged, perhaps placed in the spider hole for the media value of plucking him forth in this wretched condition. Even if this were the case, it seems extremely unlikely (to me, anyway) that military commanders would have tipped off Stephan Schwartz weeks in advance just so he could contrive a hoax—and one that hardly anybody, to my amazement, has even heard about.

Why didn't Schwartz offer this information to the relevant authorities? His comment was this: "I no longer had access into the byzantine world of intelligence. But there were several intelligence people in the audience, by the way." My guess is that he wasn't particularly eager to be hauled away for questioning as a suspected accomplice to terrorists (however absurd and inappropriate such an accusation would be). Schwartz puts it this way: he had no way to *operationalize* the information—military jargon for *act on it*. To my mind, this paradox-thwarting fact raises some extremely interesting theoretical issues. (Briefly: if your task is to remote-view the future capture of a wanted individual, and that capture is facilitated directly by the information your remote viewing provides, haven't you created a toxic loop in time and causality? Isn't this precisely the same impossibility as hoisting yourself high up into the sky just by tugging strenuously on your toes? We must defer those kinds of philosophical questions to a later chapter.) But Schwartz insists quite convincingly that had troops been able to draw upon the description provided by the remote viewers, it was so extremely specific that they would have gone directly to that single identifiable house with a palm tree at either end, the dirt road before and the river behind, and hauled Saddam out of his hidey-hole.

Why did this particular exercise in remote viewing work so exceptionally well? Because of its *numinous* quality, Stephan Schwartz argues, its intense emotional charge for the American viewers during the frenzied shock and awe and then protracted frustration of battle against the Axis of Evil. Immediately, I hear the distant but resounding cries of scornful disbelief, the screamingly obvious and shouted question everyone will raise, the inevitable challenge: "If you can do this for Saddam using a bunch of raw recruits, *WHERE'S OSAMA?*"

Possibly the answer to that question will be known by the time you finish reading this book. On the other hand, it's been more than five years now, as I write, since the murderous attacks on the World Trade Center, and Osama bin Laden has managed to elude his pursuers, despite the efforts of the world's most formidable military and an immense price on his head. Would a remote viewer have motive to attempt this task? I could think of two immediate reasons to try: First, patriotism, concern for one's fellow citizens, as capturing this figurehead would be a coup against what President Bush had dubbed Islamic fascism. And, second, there was that bonus waiting: one might make parapsychology or oneself rich at the same time with the $25 million reward.

Of course, I put this question to Schwartz. His answer was both predictable and frustrating, and raised precisely the deeper question posed by the Saddam prediction:

I have answered this question in private conversation and from public podiums dozens of times. I will do an Osama probe any time someone can assure me that the information will be operationalized in a timely manner. I did Saddam to make the point. It was irresistible. I knew Saddam's discovery, whether dead or alive, and whenever it occurred, was a powerfully numinous target. That meant there was a high probability of success implicit in making the effort. And doing it with a group most of whom had never done a Remote Viewing made it, I confess, all the more attractive. I knew going in the chance of anyone acting on the information, and was o.k. with that. But what's the point of doing this again? To prove it can be done? I've found sunken ships, Cleopatra's Palace, Marc Antony's Timonium, the remains of the Lighthouse of Pharos, what are probably the remains of one of Columbus' caravel, and some other lovely things. We've helped solve murders using Remote Viewing. And my work is just a small part of the total Remote Viewing research. Given the research on record, questions about the reality of remote viewing strike me as being based on willful ignorance. Like believing the world was created 6,000 years ago.

I've thought about doing Osama at some length, however. Technically, given the strengths and weaknesses of Remote Viewing, his location will be more difficult than Saddam's, mostly because his environment is so homogenous. His geography is more natural, there aren't villages, hospitals, cities, road complexes, the kind of geographical features a capture team would need to discern the actual location. If I told you Osama was in the mountains along the Pakistan Afghanistan border, moving with some frequency, and went on to describe a cave or house, that alone would not be enough. It requires a dedicated special operations field team, and the sort of satellite imagery they have access to, that can be used for the location component. You'd want the team to be able to communicate back additional questions [to the remote viewers], and use those answers to refine the location. But I have no doubt whatever that it could be done. (Personal communication, August 27, 2006)

Is this a satisfactory answer? There are yet more reasons to be cautious. It struck me as highly likely that the conspicuous and humiliating live capture of Osama would (will) actually lead to an *intensification* of the jihad, with still more deaths as a direct result, so maybe the game would not be worth the candle. He might be replaced by someone even more ferociously dangerous. On top of that, there was always the risk that some official idiot would regard information obtained by remote viewing as proof that the psychics must be in cahoots with Al-Qaeda and slam the lot of them into Guantanamo Bay. Worse still, if the remote viewing exercise proved successful but official ears remained adamantly deaf, there might be a temptation to announce the findings prematurely in public, thereby cruelly compromising and possibly aborting an ongoing search. It turns out, anyway, that numerous self-described "remote viewers" have been posting on the Internet what they claim to be information on this vexed topic. Since there's no way to assess their skill or standing in this murky area, of course nobody takes these predictions any more seriously than rational people took the chatter about a gigantic UFO tearing along behind the comet Hale-Bopp.

Which brings us to a skeptical question even more piercing than the one about Osama bin Laden, and that is: *Why the hell should anybody believe a word of this preposterous story?* Was Schwartz's remote viewing forecast of Saddam's capture published in a reputable scientific journal? *Science,* or *Nature?* This, though, is a self-evidently bogus test, since those journals would either bluntly refuse to publish such a communication, or do so only if it came under the seal of, say, the U.S. president's science adviser. Very well, was it at least published in one of the peer-reviewed journals of parapsychology, hardly a credential in the eyes of skeptics, but evidence of a kind? Alas, no. The article Stephan Schwartz links to from his site is a five-column illustrated piece from *Venture Inward,* a bimonthly journal published by the Association for Research and Enlightenment. ARE is a sister organization of Atlantic University, which has its own charter and board. Perhaps you are unfamiliar with this institution of higher learning? A quick educational diversion, then . . .

⊙⊘

Edgar Cayce (1877–1945) was a clairvoyant and prophet in the early decades of the twentieth century, an American seer living in Virginia Beach, Virginia, who communed with spirits and dumbfounded his acolytes with such interesting forecasts as the convulsive rise of the lost continent of Atlantis from the deeps, to be expected confidently sometime toward the end of the twentieth century, although probably not before the conversion of China to Christianity by 1968.

> The Great Lakes would empty into the Gulf of Mexico linked with a time when ancient repositories would be discovered as people reached the appropriate level of consciousness. The three repositories mentioned are Egypt, the Bimini area, and the Yucatan.
>
> Activities by Mt. Vesuvius or Mt. Pelee, or in the southern coast of California and the areas between Great Salt Lake and the southern portions of Nevada, we may expect, within the three months following same, inundation by earthquakes, more in the Southern than the Northern Hemisphere. Portions of New York, or New York City itself, will disappear as well as the southern portions of Carolina, Georgia.
>
> Land will appear in the Atlantic and Pacific.[2]

Cayce channeled some thirty thousand medical diagnoses and provided recommendations for "natural" treatment that probably were no worse than the pre-antibiotic, pre-hi-tech conventional medical wizardry of his day. Today his legacy is maintained by the Association for Research and Enlightenment, which supports the teaching of his methods through Atlantic University, an accredited distance-learning institution (you can study there on the GI Bill) also located in Virginia Beach, offering MA degrees in holistic studies. You can build up credit with courses along these lines:

LP501p—Listening and Dialogue

October 27–29, 2006 (with follow-up distance learning modules)

Instructors: Amy Betit and Tom Curley

This course will begin with an exploration of the inner depths of your own experience, through silence, listening to nature, and listening to self. Methods of listening and dialoguing with others will then be practiced. As part of the skills you will learn as a facilitator of group dialogue, you will create your own talking stick, for use in future group situations. This course will be held in Pungoteague, VA. Enjoy this unique opportunity to spend time in a peaceful, rural setting to examine the many aspects of the art of listening.

You might be tempted to roll your eyes at this course description, but I assure you that it can't possibly be any more useless than the blithering word-salad nonsense I had to put up with at the deconstruction seminars I attended mutinously for some years during my PhD studies back in the late 1980s.

Schwartz notes: "What Edgar Cayce said half a century ago has nothing to do with my research, except both involve attempts to access nonlocal awareness." Yet it's easy to imagine skeptical rationalists having their doubts about this remarkable story. Even if they grant the probity of Atlantic University and its administrator (at least it's not called Atlantis University), how can anyone be sure, they'll ask, that the documents allegedly produced by these forty-seven participants really were created in advance of Saddam's capture? How can anyone be sure that they exist, for that matter? We have been told that the originals are locked away for safekeeping, which means nobody can look at them to determine the validity of the claim.

I find this objection specious. The documents have been notarized —a legal guarantee; the students themselves know very well what happened, and three years later none of them has come forward to deny Schwartz's account. But wait—*did they even exist?* Schwartz told me:

> These people paid money to attend this seminar. They signed up. They gave addresses, telephone and credit card numbers. There is an entire paper trail on each of these people. Their voices appear on the tapes. They will one by one tell you all about this should anyone be willing to invest the time and energy to track them all down. Frankly, I'd love to have someone do it. A massive amount of work, but it would make the record even more complete. I design these things like fractals. I don't care how deep you want to go, it'll be the same result all the way down, because this is actually what happened, and it can be validated by anyone from a dozen directions.

I contacted administrator H. A. Stokely and requested confirmation of this account. He gave it willingly:

> The event in question was a conference on Remote Viewing hosted by ARE. . . . The course portion involved about two and a half days after the conference where a smaller group studied and worked with the material in more depth. . . . During the course, a practical exercise was proposed. The consensus that emerged was to view the circumstances

surrounding the finding of Saddam Hussein. The session was conducted by Stephan and the results—mostly pen or pencil drawings on plain white paper—were collected by me. . . . At Stephan's request, I quickly made him a Xerox copy of each sheet, sealed the originals in an envelope—and had a notary view my signature across its seal and date its sealing. The envelope was then placed in the vault. . . . To my knowledge it has not been removed since that time. (Personal communication, September 2, 2006)

The editor of *Venture Inward,* Kevin J. Todeschi, confirmed as well that he knew in advance the details of the consensus account:

Those conferees did a remote viewing exercise and collectively obtained some information that essentially "foresaw" where Saddam was. Stephan talked to me about this exercise and the information. . . . I remember this because one of our conferees got *very angry* that a "spiritual conference" would have as its focus this kind of project—something the individual thought was entirely *not* spiritual. . . . I am happy to confirm that I do recall Stephan telling me after the conference in November that Saddam would be found in an underground cave. (Personal communications, 8 and 11 September, 2006)

The best and only riposte to this corroboration I can think of, based on the premise of sheer random or inferential coincidence of detail (which each individual must assess), is that hundreds if not thousands if not tens of thousands of wannabe RemoteViewers@home might have attempted this same impossible task, and the only ones left standing were those who managed to get close just by guess or by damn.

That explanation doesn't hold water, either. We're not talking here about an unsubstantiated claim made by some fruit loop plucked from the Internet. This is a documented procedure conducted by one of perhaps only two or three handfuls of people on the planet who are recognized by their parapsychological peers as long-time experts in this field. I'm not saying that if Josephine Homebody and her husband and seven children had produced such a consensus report it would necessarily be fraudulent, or the result of chance—but I am saying that there'd be materially less reason to take it seriously at face value.

What about the "file drawer" problem for the individual investigator, a supposed surfeit of nonsignificant studies hidden away and never published? Schwartz declares that his statistics are comparable to those of other remote viewing experimenters: "My experience in these applied high numinosity, high entropy target projects is pretty much what has been reported since the '70s by me and everyone else who knows what they are doing. By concept analysis, typically 75–85 per cent of the material will be evaluated by independent judges to be correct or partially correct." What's more, "Out of the 50 or so multiple viewer applied experiments I have done, I have had two notable failures."

Perhaps it needs to be stressed that the Saddam prediction is not the only success story from the files of remote viewing. Stephan Schwartz mentioned several that are equally astonishing, but not yet written up or published: Predicting the Challenger shuttle disaster—after NASA engineers had expressed their concerns—and pinpointing the problem with the O-rings. Identifying significant details concerning the spree-killer Washington Beltway snipers, John Allen Muhammad and Lee Boyd Malvo, in 2002. And one multi-viewer documentation I have had the opportunity to read is the case of Susan Smith, the distraught South Carolina mother who drowned her two small boys by driving her car into John D. Long Lake in 1994. (This was perhaps a peculiarly numinous site, if not to the same degree as the previous pair, because two years later three adults and four children also drowned at the same place *while visiting the spot of the Smith infant deaths.* Unbelievably, their runaway vehicle knocked down a sapling that had been planted as a memorial of the Smith children.) At the time (October 30–November 1), nobody knew where the children were; Smith, although under suspicion herself, claimed for nine days that they had been abducted in her missing vehicle by a black carjacker.

One of Schwartz's viewers, who knew nothing about the case, because he had been too pressed by work to watch the news or read papers, stated: "They are dead. . . . There is something wrong with the mother. . . . I get the sense that she is terribly ashamed. . . . I sense water. . . . I want to say a watery grave. I see them in a car, underwater." Another, independently, said: "Amorphous muddy wet place. . . . Feel I am in some kind of box . . . you know maybe it's a car. Dirty. Muddy. Damp. It's like it's underwater. . . . I don't feel good about this. This is hard to do." Which is a reminder that the remote viewing of disasters is not a game. These are sensitive people, putting themselves into situations of horror. A third viewer said: "It's dark and. . . . watery. But they are inside of something . . . metal . . . like a big metal box." Various details were provided about the possible location.

On November 3, 1994, Susan Smith was still saying on *CBS This Morning*, "I did not have anything to do with the abduction of my children." She added: "Whoever did this is a sick and emotionally unstable person." Later that day, confronted by Sheriff Howard Wells, she broke down and asked for a gun to kill herself, then wrote a two-page confession confirming in most respects the remote viewers' perceptions. After her trial, she was sentenced to thirty years in prison, due for release in 2025 at the age of fifty-three.[3]

Here is Stephan Schwartz's consolidation of those several projections, made days in advance of the discovery of the little boys' bodies:

Analysis:
1) The children are dead.
2) The mother is involved and perhaps a principal in their death.
3) The mother is psychologically disturbed and was estranged from the children's father.
4) The mother's account is a fabrication.
5) The media accounts are all wrong.
6) The father was not involved with these events.
7) Another man may be involved with the mother, but there is no indication he was involved in the events that led up to their death.
8) The bodies are in or near a park, not far from where the family lived.
9) There is a strong probability that the children are in a car or metal box, in a body of water in or near the park.
10) No black man is involved with events surrounding the death of these children.

Would the police or the district attorney have been able to make use of this information had it been presented to them, as happens routinely in the heavily fictionalized television drama *Medium?* Almost certainly not, although Schwartz also provides ample descriptions of cases where police forces *have* successfully made use of remote viewing results. The problem here is that details are often kept confidential, so it's impossible for an outsider to test the claims. One that I would love to record in detail, except that Stephan is bound to silence about the specifics, is the case in which, he tells me,

The stake-out of a dentist resulted in the arrest of the perpetrator. The stake-out location was based entirely on the viewers' description of his [future!] arrest. Indeed, no one even knew this dentist had a connection with the perpetrator, until the viewers described his previous visits there, correctly after noting that the murderer had bad teeth. They described when he would next be there for treatment.... We could not get the dentist's name, but I had the viewers describe his waiting room, and the police then visited all the dentists—six or seven were in the town—until they found artwork on the walls that matched the RV descriptions. On that basis they staked out the office, on the day predicted, and the suspect showed up.

This case has never been published, and was undertaken under an agreement of confidentiality for a variety of reasons, not least the detectives involved were afraid about anyone finding out what they had done. I agreed to do this at the behest of the husband, who was stricken over his wife's murder. The killer was a very bad guy. Perhaps some day I can say more about this, but not now.

Even more compelling, because easier to test, are the public exploits of Joe McMoneagle. This bluff ex-soldier, a veteran of Vietnam, has continued his remote viewing tasks under the gaze of television cameras on numerous occasions, notching up a sequence of dazzling successes, and only a few duds—most of those plausibly attributable to botches in the protocol caused by hasty or careless TV crews. The peculiar nature of the response to these exploits is difficult to convey except by analogy. Suppose that in 1954, a young British medical student named Roger Bannister had decided to prove to the world that the four-minute

mile was not an unbreakable barrier. It has been known for decades
that nobody can run fast enough to come in under four minutes, just
as no jet aircraft can possibly break the sound barrier. The laws of
nature and human feebleness foreclose on that option. Bannister,
deluded idiot that he is, takes off down the straight and runs the first
three quarters under three minutes. People begin to look at their
watches with amazement. Is it conceivable that he will—? Bannister's
legs pound, grass flies up from his cleats. He snaps the finish line tape.
Judges consult their chronometers—3 minutes 59.4 seconds. Bannister
has collapsed. The crowd is cheering. But what's this? The organizers
are shaking their heads. One of them is consulting a Bible, another is
checking the works of Einstein. It is clearly impossible for a man to
have performed this feat. The crowd is uneasy, but they know authority
figures when they see them. Explanations are sought for the mass illusion. A tremendous tornado must have gusted behind Bannister,
hurtling him forward. No, it was an aberration of the timepieces, produced by the amount of sweat lingering in the air from the previous
race. Or conceivably, some kind of optical effect misled the observers.
The authorities huddle together. Rumors begin to spread that the fix
was in, that the runner had bribed the judges. This outrageous suggestion is quickly repudiated, but the authorities on running finally send
their spokesperson to the microphone. Sorry, folks, it might have
looked impressive to the lay observer, but what happened here today
was nothing more than a coincidence. The four-minute barrier remains
unbroken, and will until the end of time.

You think I'm exaggerating. Things couldn't possibly have dete-
riorated to that extent. Well, let's see. In February 2005 National
Geographic decided to produce a program on paranormal claims as
part of their *Naked Science* series. A negative spin seemed likely, but
a number of specialists in apparent anomalies gritted their teeth
and agreed to participate. After all, National Geographic was one of
the popular hallmarks for integrity, noted for bringing unusual

knowledge to the masses. Joseph McMoneagle and Edwin May were invited to take part in this hour-long program, and somewhat reluctantly agreed to do so, as long as the protocols were specified in advance and followed exactly. The television producers agreed. In the event, McMoneagle and May agree that it was an exceptionally well-controlled trial of remote viewing. Still, you probably think this is going to turn out badly. You would be astute in thinking so.

I watched this program with gathering surprise, as one after another apparent psi effect was demonstrated—then viewed the scoffing conclusion with a sinking and incredulous heart. Joe McMoneagle, in his blog for July 30, 2006, provides a stinging summary of what went down, and here I paraphrase his account. The producers hired a specialist to select target sites inside the San Francisco Bay area, a woman unknown to McMoneagle but instructed for several hours in good target selection criteria by May, staying in a different town and with no further contact until the trial run was concluded. Locations were selected and photographed, and the images and identity of the alternative sites were sequestered by a law firm hired by National Geographic, not known to the paranormal team. Not until the day of the shoot was the target selected from these options by somebody at the law firm, then taken to the filming location by a policeman hired by National Geographic. He delivered a sealed envelope containing the identity of the target site, a second sealed envelope holding pictures of the target and five (not, as he says, four) other mutually distinctive decoys, and a third envelope with a photograph of a beacon person who would go to the site. All of these images were unmarked by fingerprints or other identifiable elements.

The beacon person, or Outbounder, traveled straight from the law firm to the random site. With cameras rolling, Dr. May took two of the sealed envelopes—one with a photograph of the beacon, and the other, which remained sealed, with the possible targets. Opening the picture of the beacon, he showed it to McMoneagle, established for the camera that Joe had never seen the woman before, and asked for a description of where the woman was now standing. Within half a minute, Joe McMoneagle swiftly drew an image and described it thus: "Circle of dirt or gravel, with an art form in the middle. Flat." McMoneagle adds: "It was done so fast in fact, the National Geographic observer was extremely disappointed. So, I then said; 'Gee, I guess that's not much for a National level television show, is it.' I thought for a moment about the target. Then said; 'So, why don't I describe exactly how the Outbounder got to the targeted site,' which I then proceeded to do."

Still with cameras rolling, McMoneagle drew arches and the entry into the area where the beacon was standing, then was taken to a van so that he might be driven to the scene. Once he was out of the room, Dr. May was given the sealed packet containing the photos of the six possible targets. He had not been permitted, of course, to open the sealed photograph of the actual randomly selected target. Inside a minute, he matched Joe's descriptions with one of the images—that of the correct target, for a first-placed match. May stated on film that it was quite clear from the drawings which of the six was the target.

Meanwhile, the producer phoned the beacon person and obtained her location, asking McMoneagle to take the front seat in the van so that their arrival would be filmed. Note that only the driver and the producer knew the destination at this point. When they arrived at the site, they found two police officers, one a California Highway Patrol officer assigned to Home Security for Anti-Terrorism, which provided a certain grim amusement value. Seeing the crew photographing down along the edge of a bridge, the officers checked them out. McMoneagle comments: "Both stayed when they heard that I was doing the RV, and had called in to their superiors for permission to stay so they could meet me. Both reviewed all of the materials. Both officers said based on what they saw, they would have gone directly to the bridge based on my drawings alone."

And the reaction of the National Geographic crew and producer? They expressed amazement and baffled incredulity at what they witnessed. "It was very clear that they were convinced that it was successful, replicable, and real," McMoneagle concludes. He himself was not especially astonished. "That was my 86th live demonstration of double-blind remote viewing on national level television, and at the time I was running an approximate 88% success rate."

So, then, what was the post-event interpretation displayed on the television screen? The experts were unimpressed. All of this, they said dismissively, was undoubtedly nothing better than a coincidence. After all, Dr. May had one chance in six of selecting the target by absolute guesswork. You get the impression a chimpanzee could have done it by reaching into a hat and pulling out a number. The copious valid and surprising details were ignored. Because it couldn't happen, therefore it hadn't happened. Just as Bannister had not broken the four-minute mile, because humans just can't run that fast.

☙❧

Other apparently successful and really rather astonishing phenomena shown on the *Naked Science* program were explained away by much the same doublethink hand waving. What's particularly charming about this approach is that if the criticism is valid (success at the task means nothing, because its a priori likelihood was one in six anyway), why did producers bother with the test in the first place? One can only assume that they expected the psychics to fall flat on their face. After all, there were five ways out of six of going wrong, 83.3 percent in favor of the house. How delicious it would be to trip up Joe McMoneagle, supposed psychic superstar of the ludicrous CIA boondoggle Star Gate! When that didn't happen, the fallback position is obvious: not significant, couldn't be significant, the tests were not controlled, and anyway such an impossible phenomenon has never been replicated. This last bogus claim is the most outrageous aspect, I suppose, since of course precisely these protocols had been trialed again and again both by the Star Gate project and by a number of parapsychologists outside the program, yielding a cumulative success that even Dr. Ray Hyman was obliged to acknowledge, even if he thought it was due to some as yet unknown flaw in the methodology.

So to the paranormalists the galling aspect of the *Naked Science* show was that it repeatedly, and surely with full foreknowledge, set up demonstrations that *couldn't* (by the producers' standards, or perhaps by any reasonable standards) *prove* the reality of psi. They *chose* to test the physiological responses of nine-year old Richard Powells as his twin brother, Damien, in another room with their mother, Anna, was repeatedly subjected to startling stimuli, only to declare that the apparent repeated successes were meaningless. Why? Because sensory leakage and cuing were possible explanations—since the boys were inside a single building, although separated by closed doors and two floors. Disingenuously, the narrative voice-over asked: "Can we draw any conclusion about telepathy from the test? Probably not. For one thing, some question using a polygraph machine to test telepathy. Body function might not tell us what's going on in the brain. And then there's the question just how physically separate the twins were from each other. They were apart, but far from perfectly separated from each other . . . perhaps not far enough to have heard something or felt vibrations from downstairs."

Yet these were *the very conditions the filmmakers themselves had freely chosen,* the very testing apparatus they decided in advance to use. There was also apparently deliberate deception. The program presented Joe McMoneagle's comments during the drive as he recognized one aspect after another of what he had remote-viewed, as if *that* were the validating element, and then accused him of retrofitting his comments to what the camera was observing—skidding past the fact that it was Ed May's accurate blind judging, taking place elsewhere, that really counted and was the key replication of standard RV protocols. And of course the program's overarching sin was failing to set these tests in the context of their rich statistically significant background, as if this prior did not exist, and then declaring *all* the startling and effective results, somehow and without argument, insignificant and unrepeatable.

And yet, amusingly, I think what most viewers probably take away is the program's visual concatenation of positive effects, ignoring the mandatory and predictable verbal negatives throughout and especially at the wrap-up (in part because people recall visual *yeas* much more vividly than verbal *nays*). If I didn't know better, I would be obliged to conclude that the show was a deliberate exercise in persuading viewers of the *reality* of psi while piously pretending not to.

McMoneagle has not always been so ill served by media presentation of his skills. In Japan he has been treated respectfully, protocols have been adhered to rigorously, and his results have been even more startlingly impressive. Here are some examples: In March 2006 McMoneagle made his tenth appearance on the top-ranking, longest-running TV show about the paranormal in Japanese history, *Chounouryoku Sousakan* (*FBI: Psychic Investigators*). At the time, he was in considerable, almost disabling pain from back injuries suffered in Southeast Asia nearly thirty years earlier in a helicopter accident. Within months, he was scheduled to undergo four and a half hours of surgery—lumbar laminectomy, spinal-stabilizing titanium screws and rods, disk fusions, pain-control-device rewired. I mention this simply as an index of how much distraction a skilled remote viewer can withstand while searching, under the stress of the camera, into the unknown for information. The television program began as usual in Virginia, where McMoneagle spent two full days remote-viewing the location of three missing Japanese people, freely responding in the absence of any feedback or advance knowledge ("front-loading"), guided only by sealed opaque envelopes containing names and birth dates, if available, of the missing. The show would be broadcast live in about four weeks, giving the producers very little time to use this information in their attempt to locate the missing people. They used their time well.

McMoneagle and his wife flew to Tokyo, where it was announced on the show that (in his wife Scooter's words):

> As a result of Joe's remote viewing information, detailed maps and drawings of landmarks, they were able to locate a woman who had been missing for a period exceeding seven years. She was found in a small city to the north-northwest of Osaka, approximately 200 kilometers from the town from which she had disappeared. She has since been reunited with her family and friends. On the show, it was formally announced that Joe had at that point searched for 26 missing individuals and was successful in finding 13, some of whom had been missing as long as 60+ years who even the police and private investigators had been unable to locate after many years of searching.

In mid-2004 the same blind procedure led to the location of a misplaced fifty-eight-year-old Japanese mother who, because of ill health, had given up her daughter more than thirty-four years earlier. Into the bargain, McMoneagle found the sixty-year-old father as well. The parents had separated at the time the child was given up, but both were located in different sectors of the same city on the island of Hokkaido. The oddity (I mean, yes, the whole thing is outrageously odd, but this is odder still) is that McMoneagle reportedly confused the parents' houses, attributing each to the other. Scooter McMoneagle states: "The film crew was able to identify the right city by the outline of the island Joe drew and through the specific 'photograph-like' pictures Joe drew of the train station, the main bridge, primary road layouts, primary mountain features, main city feature layouts, such as the position of the larger hotel, subway tracks, river, river walkway, main shopping area, etc." In an emotional television occasion, mother and daughter were reunited by phone, to protect the parents' privacy.

Earlier attempts to locate missing Japanese are described in *The Stargate Chronicles*, where an equally remarkable saga is told about Mr. Noriyuki Ito's search for his mother, who had also abandoned him when he was only a year old, and had not been seen for twenty-seven years. Again, the remote viewer did not even know this much; he was told only: "Describe the person and their current health or condition. Describe the location of the person." Not much to go on—not even gender. On film, McMoneagle generated information that specified a living middle-aged woman, resident in a delta between a river and the center of a city with a bay and four constructed islands, near a large parklike area, a round monument-like structure in the city's center, the relevant prefecture not exceeding three hundred thousand in population. She would be living in the back-most room of a three-story apartment containing six rooms. Drawing on this glut of information, first scratching possibilities and then homing in on the most likely choice, the TV research crew found a three-story apartment building within the relevant delta—with, granted, twelve rooms rather than six. The mailboxes showed that the back-most room of the first floor *bore the maiden name of Mr. Ito's mother*. A man was at home, but fled the TV crew. Checking the records at city hall, they learned that this room was indeed the woman's residence. In 2002, when McMoneagle's book came out, Mr. Ito had not yet received any response to a letter he wrote to his mother at that address. A frustrating outcome, but hardly one especially congenial to a stringent skeptic, I think.

᎒᎒

So here's the inevitable question: if all of this eye-popping stuff is not moonshine and lies, if psi is now validated and can fetch us information not otherwise available, *why the hell did the U.S. government close down its pioneering remote viewing program?* What possible motives could they have? How did they *justify* such a rash act?

Actually, it is now known that the United States government was not alone in exploring remote viewing. In February 2007, following requests made under the Freedom of Information Act, the British government released a package of heavily blacked-out documents, formerly stamped Secret-UK Eyes Only, on remote viewing experiments its military had trialed from early November to late December 2001, not long after the September 11 attacks. The Ministry states that these experiments were meant to "investigate theories about capabilities to gather information remotely about what people may be seeing and to determine the potential value, if any, of such theories to Defence."[4] The timing hardly seems coincidental. The documents reveal a rather ill-prepared essay into the paranormal, and close by reporting that preliminary attempts were insufficiently rewarding, so the short-lived program was closed down. Despite the deleted data, a somewhat different story can be read in these odd pages.

The British Ministry of Defence paid a paltry £18,000 for this set of initial exploratory tests, which were intended as a sort of dry run at the real thing. To start with, the MoD officials hoped to attract the cooperation of twelve psychics claiming on the Internet to have learned how to do remote viewing—but to their chagrin none of those allegedly gifted individuals were prepared to work with the military. So the psychologists in charge of the program found naive or novice subjects, and the psychologists themselves took part in several tests; two of them withdrew after a single test and declined to be tested again, although their motives for refusing have not been disclosed.

Twenty tests were attempted—but these were by no means comparable with most of the impressive long-distance trials routinely recorded by US researchers. Targets were plucked randomly from sets containing six pictures, each distinctive, with the would-be viewers using a protocol adapted from online documents thought to have been designed by American Ingo Swann. Inside a small house rented for this purpose, the viewers were blindfolded and asked to describe and draw images of the picture masked by an opaque envelope. Various electronic sensors were deployed to see if their nervous systems were doing anything noteworthy and illuminating during successful viewing. This sounds more like a somewhat fraught version of free response ganzfeld than Star Gate-style remote viewing, and the long training sessions that we are told honed the skills of a Joe McMoneagle were not undertaken during this very abbreviated period. This was intentionally part of a larger project, with these results meant to form a kind of baseline against which the scores of true sensitives might be fitted, assuming real psychics could ever be attracted into the research program.

Most of the targets, and all of the decoy non-targets, remain classified; perhaps they related to a search for terrorists, supposed weapons of mass destruction in Iraq, or some other touchy matter. Known targets include a portrait of Mother Teresa, a partially open pen-knife, and an empty gas station. Of the twenty trials, two were deleted entirely. These should be removed, therefore, from any assessment. In fact, the MoD report speaks only of eighteen tests. But one of the remaining eighteen was a "no score" because the viewer could not perform, and another was interrupted (by an electricity meter reader!). Counting the latter but not the former leaves seventeen plausible tests with assessments. Of these, the majority are summarized as revealing no access to their targets (that is, they were entirely wrong), *but six indicate at least some access.* One subject succeeded partially on three occasions, another twice, and a third once.

Apparently no formal ranking of calls against targets was done, no statistical analysis. And yet, in brief, these really rather poorly done tests with untrained subjects scored with 35.3 percent success, of a kind, more than double the mean chance expectation of 16.7 percent (that is, one opportunity in six of success purely by chance in each test). That result compares favorably with most ganzfeld experiments. Did this lead immediately to stricter and more elaborate testing? No, we are informed. The Ministry of Defence concluded that the theory of remote viewing was of little value and the program was discontinued. This finding would seem to be on a par with pulling in twenty people off the street at random to check on whether, with no training, they could all run a Roger Bannister four minute mile, or paint like Rembrandt. So preposterous is that image that it is difficult to believe further investigation will not eventually turn up more competently conducted experiments in remote viewing by the UK military.

To repeat, then: how could authorities justify such rash acts?

4.

SEEING THE LIONS

Some "advanced training" targets were non-military, and were instead chosen for their maximum extrasensory impact on the viewer—a fifteenth-century brawl in Cornwall, the surface of an asteroid, the nuclear reactor meltdown at Chernobyl, a vertigo-inducing spot a mile above Grand Canyon. One time, Dames gave Riley a target, and Riley had a bilocation-type episode for a short period. He was in a kind of hypnotic, deep-*ERV* [*extended remote viewing*] trance, murmuring descriptions of what he saw. He saw a stadium, and oddly dressed people going in to see a football game. He went inside the stadium, and a roar rose up from around him, shrieking, boisterous, a familiar sound, but there was no football being played here. He saw the lions—this was the Colosseum, in Rome, circa 100 A.D.—and Dames pulled him out of there, and Riley was in something of a zombie mood afterward, bouncing off the walls.

—Jim Schnabel, *Remote Viewers*

Edwin C. May, who was born in 1940—and therefore, like me, isn't young enough to be an aging baby boomer—is currently director of the Laboratories for Fundamental Research in Palo Alto, California.

He was awarded a science degree in physics in 1962 from the University of Rochester, and in 1968 a PhD in physics from the University of Pittsburgh. (His dissertation, in case you're wondering, was "Nuclear Reaction Studies via the (p, pn) Reaction on Light Nuclei and the (d,pn) Reaction on Medium to Heavy Nuclei.") After three years of postdoctoral physics at the University of California–Davis, he pursued a number of subsidiary careers, as a physics instructor and in computing software and hardware, before becoming a consultant and then senior research physicist in the Psychoenergetics Program at SRI from 1976 to 1985; program manager of the Cognitive Sciences Laboratory, SRI International, to 1990; and then director of the Cognitive Sciences Laboratory when it moved to Science Applications International Corporation (SAIC), also in Menlo Park, until 1995. Behind this apparently straightforward résumé lurks the fact that May was in charge of the research end of the top-secret U.S. government remote viewing program, until it was ignominiously shut down in 1995.

That made Ed May perhaps the most informed and influential player in the world of psi in the . . . well, in the entire world. At any rate, in the United States. Other experts in the paranormal—or, as Ed prefers, *anomalous phenomena*—have been working industriously all these years, especially in notable but not particularly well-funded laboratories such as Edinburgh University's Arthur Koestler–endowed chair. Then there's Dr. Harald Atmanspacher and his colleagues at the Institut für Grenzgebiete der Psychologie und Psychohygeine in Freiburg; Professor Deborah Delanoy at the University of Northampton; Dr. Dick Bierman's chair of parapsychology in the Netherlands, formerly at Utrecht University, now at the Humanistic University (also in Utrecht)—not to mention various establishments in Sweden, India, and other countries—and well-placed independent parapsychological scholars like Stephan Schwartz. But Ed May was right at the center of the action for years, even though the rest of us were not allowed to know what he was doing, mostly. Fortunately, that has changed, to a large extent, after the declassification of Star Gate documents and records.

The program Ed May ran, and before him Hal Puthoff and Russell Targ, was funded on an annually replenished basis according to the success of their taskings, as evaluated by the armed services and intelligence community who listed them in their appropriations. The figure generally given for this investment is on the order of $20 million over roughly twenty years—small beer by military standards, true, but not the sort of money that would be thrown away for decades on foolishness, at least not without serious testing of the claims and findings of the research organization. In March 1996, in the *Journal of Parapsychology,* Dr. May published a remarkable history and assessment of the review of Star Gate that resulted in the program's closure. A similar overview was offered by the program's first director, Dr. Puthoff in his "CIA-Initiated Remote Viewing at Stanford Research Institute" (*Journal of Scientific Exploration,* Vol. 10, No. 1). Taken together, these documents help explain why a project of such promise was allowed to peter out before being knocked on the head. May opens with a brisk summary:

As part of the fiscal year 1995 defense appropriations bill, responsibility for the government-sponsored investigation and use of ESP [his footnote adds: hereafter I use the term anomalous cognition (AC) instead of ESP] was transferred to the Central Intelligence Organization (CIA). In a Congressionally Directed Action, the CIA was instructed to conduct a retrospective review of the 24-year program, now known as Star Gate, that resided primarily within the intelligence community. The analysis was to include the research that was conducted since 1972 at SRI International and later at [SAIC]. In addition, the CIA was to include an assessment of the intelligence-gathering utility of anomalous cognition

(AC), and the program history was to be declassified. Initiated in June 1995, the evaluation was to be completed by September 1995.

That assessment would be done by the American Institutes for Research (AIR), and draw upon the expertise of the same Dr. Ray Hyman who had chaired the parapsychology subcommittee for the National Research Council (NRC) blue ribbon study of psi in the military context, published in 1988. Nobody doubted his scientific credentials, but it was known that Hyman remained highly skeptical of the very existence of psi. A second expert called in was statistician Dr. Jessica Utts; in the stinging rebuttal by parapsychologists to the NRC report, Utts had declared herself persuaded of psi's authenticity. Several AIR personnel were involved, and the review was coordinated by AIR president David Goslin.

As noted, the team had a mere three months to sort through and evaluate a staggering amount of research and operational material. Well, that is, they *would have* had they bothered to consult all the operational material—boxes and boxes of it—that had been opened to them. Almost all of it had been classified until then, naturally, so this prestigious group were denying themselves access to the most critical resource one would expect to be studied closely. Amazingly, neither did the assessors approach many of the most relevant participants of the program. Additionally, they were instructed to ignore the previous work and conclusions from earlier studies, such as the many meta-analyses of psi conducted over the years. Even so, they did draw upon the dismissive conclusions of the NRC report, although, as May notes, they knew perfectly well that the NRC investigators had no access to any of the research findings from the previous decades of work. May grants that his own exclusion from the assessment was justified, given that he was a principal in the matter under investigation, but it's obvious that he was furious to find his special expertise pretty much repudiated. "As a result of AIR negligence, their report contains numerous errors of fact and errors of assumption."

Since then, May told me, "I have arranged for all the research to be declassified retroactively from 1989 and from then to the end of the program." But he added, "No nuggets here." I mentioned rumors that the AIR review's dismissive report was biased because much salient data had remained classified, and asked, "Will the skeptical assessment change if everything (within asset-protection reason) is declassified?" May was outspoken in his reply:

> Totally not true. The group had access to everything; however, they looked at NOTHING! There were a large number of boxes sent to CIA from DIA [that is, the military wing of Star Gate]. I know the individual quite well who packed it all up. Years after the dust had settled and the AIR reports (Unclassified and Classified versions) were out, she and another colleague had access to the storage room in which these boxes were stored. NONE of them had been opened!! She told my other colleague to open that one and this one and in a space of about 10 minutes found incontrovertible evidence of the AIR/CIA white wash. There are a number of quite complex reasons that the program closed.

On the other hand, he warned me of many remote viewing cases that were subsequently "hyped beyond recognition by my former colleagues. One of many reasons we are out of business with the U.S. feds is too much hype by our 'friends' who grossly oversold what was possible." (May resigned from SAIC on November 28, 1995, and speaks now only as a private citizen.)

My own working assumption was that the program died because of a combination of fear of politically injurious ridicule plus an unrealistic military mind-set expectation of rapid results without being prepared to pay for long stretches of careful research to support and extend the practical methodology. May agreed, but added that intelligence requirements in the post U.S.–Soviet standoff era had shifted from strategic to tactical (think asymmetrical terrorist attacks, for example), and it seems that psi is not good at this kind of urgent tasking. As well, the champions of the program in the military were gone, so there was nobody left to "carry the ball forward." Finally, SAIC "were unsuccessful at institutionalizing the activity." McMoneagle has drawn a similar comparison between the way in which sniper expertise was repeatedly built up and let go during and after major conflicts, probably because it was regarded as immoral, even in war, until quite recently, when such expertise has been built in to the institution of arms. There seems to have been no way to train ordinary soldiers in the practice of psi, at least on the basis of information gathered by these secret programs.

So what did AIR claim to find unsatisfactory in these records, most of which they had declined to read? Hyman's report noted that SRI/SAIC claimed evidence for anomalous cognition (ESP), but not for anomalous perturbation (PK). The investigators had selected the ten most recent experiments to study, although they also drew upon some evidence from previous work and other laboratories. Hyman opens by citing Utts's positive conclusions: "Using the standards applied to any other area of science, it is concluded that psychic functioning has been well established. Arguments that these results could be due to methodological flaws in the experiments are soundly refuted. Effects of similar magnitude to those found in government-sponsored research at SRI and SAIC have been replicated at a number of laboratories across the world. Such consistency cannot be readily explained by claims of flaws or fraud." He adds: "We both agree that the SAIC experiments were free of the methodological weaknesses that plagued the early SRI research. We also agree that the SAIC experiments appear to be free of the more obvious and better known flaws that can invalidate the results of parapsychological investigations. We agree that the effect sizes reported in the SAIC experiments are too large and consistent to be dismissed as statistical flukes."

Possibly this surprised you. Perhaps you expected a more venomous denunciation of this apparently crackpot effort, at least from Hyman if not from Utts. His emphasis is more congenial. "The principal investigator [May] was not free to run the program to maximize scientific payoff. Instead, he had to do experiments and add variables to suit the desires of his sponsors. The result was an attempt to explore too many questions with too few resources. In other words, the scientific inquiry was spread too thin." The military, in short, was paying the piper, and the tune it demanded was discordant with the requirements of hard-edged science.

It makes sense, up to a point, and yet would these same military and intelligence clients have gone on paying year after year, *on this very basis,* unless what they received gave them a degree of satisfaction? Not themselves scientists, they were not interested in why the process worked, only that it did—but apparently not well enough for their purposes. Fair enough, but this need not, indeed *must* not, be misunderstood as proof that psi does not exist, or that remote viewing is bogus. Logically, all it demonstrates is that at this stage nobody knows how to use psi with pinpoint accuracy, the sort of thing demanded by an increasingly hi-tech military that can gather intelligence by snazzy conventional means that don't have the added odium of the paranormal clinging to them. (Granted, these wonderful technological methods misfired badly in the lead-up to the second Iraq invasion, when weapons of mass destruction were mysteriously located and then mysteriously found never to have existed. And, of course, conventional methods are not known to provide information from the future, the kind of thing that Stephan Schwartz was able to do with his group of amateurs in pinpointing the final location, six weeks hence, of Saddam Hussein.)

Hyman offers a variety of rebuttals to the apparently striking evidence from SRI/SAIC, but still concludes:

I want to state that I believe that the SAIC experiments as well as the contemporary ganzfeld [German for "entire field"] experiments display methodological and statistical sophistication well above previous parapsychological research. Despite better controls and careful use of statistical inference, the investigators seem to be getting significant results that do not appear to derive from the more obvious flaws of previous research [. . . This] does suggest that it might be worthwhile to allocate some resources toward seeing whether these findings can be independently replicated. If so, then it will be time to reassess if it is worth pursuing the task of determining if these effects do indeed reflect the operation of anomalous cognition.

This sort of conclusion to a stringent critique reads very oddly, and isn't the sort of thing one would expect to find at the end of a government-sponsored investigation into creationism, or the flat earth theory, or tooth fairy research. Perhaps it's intended, some might suggest, as a sop to powerful military funding entities that have embraced this crackpot belief system for private and pathological motives.

It would be rather insulting, though, to Hyman's integrity to suggest that this is a complete and satisfactory motive for his let-them-down-easy, soft bottom line. Indeed, aside from the somewhat noxious blend of military tasking and scientific exploration, what Hyman found most deficient was *theory:*

Without a positive theory of anomalous cognition, we cannot say that these effects are due to a single cause, let alone claim they reflect anomalous cognition. We do not yet know how replicable these results will be, especially in terms of showing consistent relations to other variables. The investigators report findings that they believe show that the degree of anomalous cognition varies with target entropy and the "bandwidth" of the target set. These findings are preliminary and only suggestive at this time.

That seems fair, and most of the rest of this book will be devoted to looking at attempts to explain psi both within and beyond the boundaries of known science.

Other objections raised by critics of SRI/SAIC results are far more problematic. One of the most famous declassified remote viewing exploits from the 1970s was Pat Price's description of a huge crane at a Soviet site in Semipalatinsk, to which he was directed by nothing more than coordinated numbers. Accounts have been given in numerous books, complete with diagrams of this immense and unusual object (e.g., in a CIA report published as an appendix in McMoneagle's *The Stargate Chronicles*). It is the kind of remote viewing outcome that simply blows the mind. Not, however, necessarily the data-hungry military mind. As Utts notes:

Although some of the information in these examples was verified to be highly accurate, the *evaluation* of operational work remains difficult, in part because there is no chance baseline for comparison (as there is in controlled experiments) and in part because of differing expectations of different evaluators. For example, a government official who reviewed the Semipalatinsk work concluded that there was no way the remote viewer could have drawn the large gantry crane unless "he actually saw it through remote viewing, or he was informed of what to draw by someone knowledgeable of [the site]." Yet that same analyst concluded that "the remote viewing of [the site] by subject S1 proved to be unsuccessful" because "the only positive evidence of the rail-mounted gantry crane was far outweighed by the large amount of negative evidence noted in the body of this analysis." In other words, the analyst had the expectation that in order to be "successful" a remote viewing should contain accurate information only.

This analyst's conclusion, then, is like one of those terrible jokes where the punch line is, "Yeah, yeah, but what have you done for me *lately?*"

The observed phenomena, and all the apparatus surrounding their acquisition and analysis, tended to flummox those who were delegated to oversee the program. So far we have heard from advocates of the program, often those directly responsible for its creation and management, and from hard-core skeptics eager to debunk it. But what was the reaction of those independent onlookers required by law and custom to keep an eye on expenditure and the direction of military/intelligence research, no matter how odd it looked? An extraordinary stereoscopic portrait became available when a secret report was declassified, and then, more than twenty years later, when the author provided his afterthoughts. First published in the highly restricted Winter 1977 issue of the CIA in-house journal *Studies in Intelligence,* "Parapsychology in Intelligence" was a personal review by Dr. Kenneth A. Kress, a senior CIA analyst whose job was to review the program that became Star Gate. This document and its afterword were published in 1999, in the *Journal of Scientific Exploration.* In his new abstract, Kress managed to place a bet either way: "Although nothing that would meet the rigorous test of science was ever achieved, there were tantalizing events and experiences that suggest the possibility of acute perceptions, either elicitation or parapsychological, in some individuals." *Elicitation?* A code word, as we'll see, for scamming the unwary.

I recommend reading the Kress report in full (it's available online, as are the papers by Puthoff, May, Utts, and Hyman) to gain a cumulative sense of the way the program rose and slipped into its long fall. It was begun in 1972, when the Office of Scientific Intelligence group (OSI), concerned by news of Soviet investigations into PK, learned of a paper by Hal Puthoff and Russell Targ suggesting that such paranormal events might be worth some serious exploration. The Office of Research and Development (ORD) and the CIA Office of Technical Service (OTS, which had been following Soviet claims from as long ago as 1961) also expressed interest. That October, Dr. Kress became the project officer; his physics background would allow him to work closely with the SRI physicists. Kress asked Price to confirm his remarkable but incomplete remote viewing reconnaissance of the Soviet structures. Because Price failed to detect four large derricks known to be in the area, the ORD officers "concluded that since there were no control experiments to compare with, the data were nothing but lucky guessing."

This dismissive assessment disturbed Kress, who thereupon began to doubt his own objectivity. Was he being drawn into a circle of subjective mutual confirmation? He asked a colleague to examine the CIA data and was told, presumably to his relief, of the colleague's "inescapable conclusion that extrasensory perception does exist as a real phenomenon, albeit characterized by rarity and lack of reliability." Still, despite some additional notably accurate remote viewing, Kress noted in his 1977 paper that during the previous two years there had been "only modest CIA and Intelligence Community Staff interest in parapsychology. . . . [D]uring November of 1976, [CIA] Director [and future U.S. president] George Bush became aware that official Soviets were visiting and questioning Puthoff and Targ at SRI.

... Mr. Bush requested and received a briefing on CIA's investigations into parapsychology. Before there was any official reaction, he left the agency." You have to wonder if history perhaps took a strange turn at that moment.

While all of this is fascinating enough, Kress's 1999 postscript is riveting. It's evident that with the passage of two decades, Kress has fallen prey to (or, alternatively, benefited from) a very common psychological process of *normalization,* an inverse form of what Hyman blames for a lot of belief in psi, *subjective validation.* The process works like this (it's probably happened to you sometime): it happened, but it couldn't have happened, therefore it didn't happen, probably. The Roger Bannister effect. Ruefully, Kress comments: "While on this quest, I did observe one effect I wish to highlight now. It was a demonstrable fact that psychics could convince professional intelligence operators of the genuineness of their powers. . . . [T]he motives of psychics clashed with mine and taught me, too late, to be very wary of psychics." In other words, psychics are great con artists, or at any rate charismatic and persuasive. Kress therefore regards this part of his career only with *guarded* satisfaction. He concludes:

> The project was fascinating and frustrating, in equal proportions. The times demanded a measured investigation that I helped to organize and manage. Others followed the CIA program with results that I will leave to history to judge. Me, I remain a skeptical agnostic. More skeptical as time advances, but careful to note that even if paranormal phenomena are entirely bogus, some individuals are surely able to instill the belief in unexplained capabilities. How they do this and what are the vulnerabilities to such enticements is worth knowing.

Decades ago, there was a wonderful comic strip called "Mandrake the Magician." Mandrake wore formal tails, a gleaming white shirt and bow tie, and top hat, as great conjurers at the turn of the twentieth century did, and he achieved his fabulous effects by hypnosis. The tag line was: "Mandrake gestured hypnotically." The master magician held forth two or three gloved, outstretched fingers, fixed his foe with a beady eye, and instantly (perhaps in part because of telepathic powers Mandrake had obtained in Tibet as a child) his opponents were flung into colorful or terrifying episodes of delusion and bafflement. You have the feeling that Kress suspects the SRI psychics of gesturing hypnotically whenever he was around.

Dr. Puthoff, not surprisingly, maintained quite a different opinion. In his review essay titled "CIA-Initiated Remote Viewing at Stanford Research Institute," he notes that although the AIR study was limited in scope to a fragment of the total program effort, it

resulted in a conclusion that although laboratory research showed statistically significant results, use of remote viewing in intelligence gathering was not warranted.

Regardless of one's a priori position, however, an unimpassioned observer cannot help but attest to the following fact. Despite the ambiguities inherent in the type of exploration covered in these programs, the integrated results appear to provide unequivocal evidence of a human capacity to access events remote in space and time, however falteringly, by some cognitive process not yet understood. My years of involvement as a research manager in these programs have left me with the conviction that this fact must be taken into account in any attempt to develop an unbiased picture of the structure of reality.

A picture of the structure of reality! This does not sound like a job for the marines, who more plausibly tend to impose their own preferred structure of reality, or at least to defend it from those who mean to challenge their values. Indeed, Puthoff points out that in the twenty years from 1975 to 1995, "In broad terms it can be said that much of the SRI effort was directed not so much toward developing an operational U.S. capability, but rather toward assessing the threat potential of its use against the U.S. by others." One might suppose that in the wake of the American nightmare of September 11, 2001, this tool in the armamentarium of intelligence might have been taken out of storage (assuming it was ever wasted in storage), buffed, and put back to work. But that's nothing more than speculation. Nobody I know of is prepared to offer any evidence that psi is back on the government payroll. In view of the colossal stupidity in the chaotic response to Hurricane Katrina in 2005—its devastation having been forecast anyway by weather experts—it's hard to detect evidence that a secret corps of remote viewers has the president's ear.

So the whole affair boils down to this: On the one hand, the results that experimenters like Puthoff and May were prepared to stand by in public remained comparatively modest—although escapees from the programs would fairly soon be telling tales of psychic contact with Jesus and the Buddha, Martians both on Mars and in secret underground caverns and tunnels on earth, UFO-riding Greys from Zeta Reticuli and points north, visits to lost Atlantis, visions of evil poisonous substances released from shining canisters during Desert Storm and responsible for Gulf War syndrome, and who knows what all else. On the other hand, the modest and accredited results of these programs were insufficiently finely tuned or accurate to satisfy the military. This despite U.S. president Jimmy Carter having admitted in a September 1995 speech to Atlanta college students that a plane lost in Zaire and inaccessible to spy satellites or overflights was located by a remote viewer.

Or *is* it possible, after all, that the CIA and the DIA and the NSA had let the program run down and decay in its own odium while quietly shuffling sideways into new programs, deeper and blacker and never to be revealed? But that is almost the defining example of an unfalsifiable hypothesis. Meanwhile, Puthoff had moved off in the direction of zero-point energy, and Ed May had taken his Cognitive Sciences Laboratory outside the embrace of the military intelligence complex into his own Laboratories for Fundamental Research, where he and colleagues such as S. James P. Spottiswoode continued their search for the signatures of psi, moving beyond attempts at *proof* (for Ed, as for Stephan Schwartz, that has long since been accomplished) to an understanding of *process,* as well as applications such as the search for lost archaeological sites, and pursuit of the elusive and evasive *theory* that will knot together psi and the rest of physics.

Shutting down the official government program has starved the principal researchers of funds, but so far they've managed to find sufficient independent income to keep their demilitarized programs running. So have they discovered anything of interest in more than a decade since that program was closed? Yes, indeed, some of it rather more interesting than one might wish. A former director of the Cognitive Sciences Laboratory is British scientist S. James P. Spottiswoode (he severed his affiliation with the Laboratories for Fundamental Research in 2006), whose earlier work with May and others seems firmly grounded in adherence to a stringent scientific approach to research—that is, classically reductionist, materialist, nonspiritualist—and yet whose most famous finding, announced in 1997, tended to embarrass even him, because it looked like—well, it looked like astrology.

It wasn't, in fact. Not that astrology and psi are necessarily strange bedfellows. Joe McMoneagle, who has done a lot of work as a research psychic for the Cognitive Sciences Laboratory, is married to an astrologer, Scooter McMoneagle, daughter of the late Robert Monroe, perhaps the world's most famous exponent of out-of-body experience and astral travel. On McMoneagle's Web site, you're likely to encounter such jaunty notes as,

My wife, Scooter, has a wonderful article that's just come out in *Sidney Omarr's Astrological Guide for You in 2007*. It's titled: "It's Time to Think Big, by Jove!" . . . She provides a lot of very interesting detail about Jupiter and how it affects each of the signs next year when it moves into its own sign of Sagittarius. She has contributed articles for the Omarr Annual for three years running, and they are all excellent. I believe her insight as an Astrologer is exceptional, but then I might be slightly biased.

To a rationalist like me, even given my evidence-based acceptance of many aspects of psi, that sort of offhand remark makes the hair stand up on my neck. Is he *serious?* Is this just the kind of pleasantry a husband expresses of his wife, not meant to be taken literally, along the lines of "My beloved is the most beautiful woman in the world"? Even raising the question is undoubtedly offensive. If astrological or Tarot or tea-leaf readings are imprecise enough to allow stray psi impressions to be incorporated, it *might* be worth continuing to use such debunked systems as projective screens—but why risk all the attendant hazards of random and erroneous content that come with astrology in particular? Better to start over with new methods, such as those developed for remote viewing or dream hermeneutics. (I keep expecting to see some new Amazon.com best seller: *Dark Matter Astrology: The Hidden Power of Dark Energy in Your Life!*) In any event, Spottiswoode's disturbing finding does *not* involve the beneficial or injurious influence of the planets, as charted by astrologers on the basis of broken astronomy thousands of years out of date. But it does suggest a link, of some very strange kind, between the effect size available to psi practitioners when they try to do their stuff and the orientation of our spinning planet, or at least the part they're sitting on at the time, toward the cosmic environment of deep space.

It's not the first time aspects of our daily life have been linked by science, rather than superstition, to the distant stars. That's probably worth bearing in mind. The nineteenth-century scientist-philosopher Ernst Mach argued that the familiar phenomenon of inertia—the tendency of things to continue moving once they've been given a shove—is profoundly connected to our celestial surrounds, the distant so-called fixed stars. Einstein's relativity seemed to sidestep this principle of Mach, and Newton before him, but it continues to nag and return now and then to nip at the ankles of scientists. What's more, we now know that the history of the planet, and of life in particular, has been fractured and remade many times by extraterrestrial impacts. The very fact that the sun's planetary system contains the materials of life and the heavy metals needed for technology is due to ancient supernovae whose titanic explosions created those elements and seeded the heavens with their life-giving dust. On a more dire note, we know that large impacting meteorites the size of mountains smashed into the earth's crust and oceans repeatedly, creating global havoc, destroying up to 99 percent of all extant life forms, and pushing evolution in new directions that eventually wandered to us. It is possible that gamma ray bursts of incredible ferocity have also swept the solar system from time to time, sleeting anything alive with gouts of radiation that slew and mutated any creatures that crept, flew, burrowed, or swam. None of this implies planning, either benign or harmful, and certainly no connection with the comforting regularities of the orbiting (known) planets and their relationships to a truly arbitrary configuration of stars—the constellations—that mostly have nothing to do with one another even in terms of locality. But that doesn't mean forces beyond the planet do not impact us. It's known that solar cycles of magnetic flux underlying the roughly twenty-two-year bipolar sunspot cycle are correlated with gusts of charged particles from the sun that slam into the earth, sometimes with heightened intensity. It is possible that the relationship between this turbulence deep inside the sun and its subsequent impact on the atmosphere and the weather helped drive cycles in human history.

That remains a controversial suggestion, but it might shed some light on Spottiswoode's finding.

Certainly many people have believed, during earlier versions of our own culture, and in different cultures, that seasons and phases of the moon act as modulators of the supernatural. Parapsychologists have checked to see whether fluctuations in the earth's magnetic field affect psi. Spottiswoode wondered if there was anything to this sort of notion, so he examined a large amount of anomalous cognition data to see if he could find any persistent, regular cycles in effectiveness—"a physical parameter which clearly modulated AC performance." For his pilot study, he drew upon a preexisting database of 1,468 free-response trials, searching their results for fluctuations in success rate. Might the time of day have some influence on psychic accomplishment? Beginning the day's work bright and early in the lab might get your participants off to a good start, or, by contrast, perhaps the drowsy relaxing time after lunch might facilitate a freeing up of the spirit. But Spottiswoode found no regular daily pattern of this kind, the wage-slave variety marked by the clock. If you are successful at nine in the morning or three in the afternoon today, it turned out not to imply that you would do quite well at nine or three every day. On the contrary. But his computer search found a remarkable anomaly within the anomalies. Like a steep mountain on the graph, there was a recurrent peak in success rates at psi tasks, but it *shifted,* day by day, little by little. Only by examining many trials did it become obvious that this constant period of high effectiveness related not to the position of the sun in the sky but rather to the distant stars. It tracked what's called *sidereal time.* The celestial clock differs from the solar clock by just one part in 365 and a bit. The sidereal day is 23 hours, 56 minutes, and 4.091 seconds long. A daily shortfall of 3 minutes and 56 seconds, near enough. Not very long from one day to the next, but in a month we're talking about a difference of nearly two hours. A full day each year. And that moving peak in success rates was nothing negligible. What Spottiswoode's results showed him, at least in this pilot study, was an improvement of *340 percent* for psychic efforts made within an hour of the Local Sidereal Time (LST) of 13.50 hours. That peak would be 1:30 p.m. if it didn't keep slipping back by four minutes each day.

This news was hard to swallow, and harder still to explain. (In fact, to this day nobody has explained it satisfactorily, although there are some sexy ideas out there.) The first order of priority was clearly to confirm the finding—or, in scientific parlance, to attempt to falsify it, for only by *failing* to *disprove* it could one have any confidence that it was a real effect. So Spottiswoode set out to find a comparable database. When he acquired one, comprising 1,015 similar trials, there was the same 13.50-hour sidereal time peak, this time still more impressively, with a success rate at the task soaring to 450 percent of the background scores.

Of course, other scientists scoffed and sneered, while new age adherents of all things mystical and holistic embraced this result with fervor. James Spottiswoode was stuck, just like Galileo staring through his primitive telescope four hundred years ago, with what he saw: "evidence of a causal connection between performance and the orientation of the receiver (i.e., a term for subject or participant), the earth and the fixed stars."

What's more, it also turned out that the effectiveness of psi *plummeted* about six hours earlier than LST 13.50 hours, and fell even more markedly some five hours after the peak hour. Granted, these data were obtained from laboratories situated in a fairly narrow band of latitudes, "nearly all the data being taken between 32 and 55 degrees North." The most unnerving element of this particular finding, apart from the sheer fact (or so it seemed) of the flux in psychic effectiveness, came when the sky and stars were examined at those hours. Most notably, the center of the Milky Way galaxy—our home galaxy—hung more or less directly overhead at the time of *least* psychic success, and was just rising above the horizon at the time of greatest success.

Is there anything special about the galactic center? Apart from the fact that a truly enormous number of stars are crowded into that compressed space, there is one special feature that caught people's attention immediately, although it was not even known half a century ago. At the very core of the galaxy, there is a ferocious collapsed object called *SgrA*—Sagittarius A star—which cosmologists and astronomers are almost certain is an enormous black hole, a sucking nothingness in space that has captured perhaps a million stars like the sun and swallowed them whole, leaving nothing but a vortex of powerful gravitation. This spinning accretion disk of blazingly hot gas and radiation is constantly reformed as material continues to slam into it, accelerating to the speed of light as it goes, and a great beam of charged particles are flung out to north and south of the galaxy by the frightful dynamics of the thing. It's the stuff of religious or at least artistic awe, looked at the right way, with imagination. Might it be the powerhouse of psi, if the LST correlation held up? No, quite the reverse. Psi seemed to be maximized when *SgrA** stood at dawn's horizon—although not, interestingly, when it was hidden by the entire bulk of the planet—and acted to suppress psi, apparently, when it was directly overhead wherever you were positioned on the earth, resembling every day a dire Star of Bethlehem, the dark side of the Force (I can't resist the religious comparisons).

In response to certain sarcastic criticism, Spottiswoode offered a crisply argued defense of the reasoning behind this search:

> The LST test was conceived for a reason which should appeal to any physicist. I do not think that psi works by the propagation of any kind of signal, but the evidence against this view is not overwhelming. In the absence of a convincing,

testable and successfully tested theory of psi, in short of a standard model, it seems wise to keep all options open.

Looking at a phenomenon that appears to involve transmission of information from one point in spacetime to another, it is a reasonable question to ask whether the effect is *isotropic*—that is, do the properties of the communication channel depend on the direction of transmission?

This is such an obvious question I was surprised that I could not think of any reference in the literature which had addressed this problem. I think there are a number of reasons for this lacuna. Firstly, most parapsychologists have been psychologists first and this question might not occur so readily. Secondly, the information source has always been problematic in anomalous cognition (AC) research, for example in ganzfeld experiments. Is the source the mind/brain of the agent looking at the target during the run? The subject's own mind/brain looking at the correct target at feedback time? The target itself at some unknown time? The experimenter selecting the target by initiating the random number generator? Or what?

So testing the effect on psi performance of the direction along which putative transmission occurred is difficult. But there seems little doubt about where the *"receiver"* is: someone, somewhere and somewhen, produces some information which is anomalously correlated to the "target." Now there is always a large object, opaque to some of the known physical fields, at the location of reception in an AC experiment: the earth. Therefore one might take a stab at the isotropy problem by asking whether AC performance depends on the direction of the vector from the earth's center through the subject at the time of reception. In spherical coordinates, the earth's rotation provides a convenient variation in *phi*, whilst a limited range of *theta* is available through the small range of latitudes of the labs at which peer-reviewed psi data has been produced. The analysis by latitude is of course confounded by the fact one is also selecting labs and therefore protocols, experimenters etc. But the *phi* component (LST) averages out all those variations, and a good test is possible.

Surprisingly, in approximately 2,200 trials from the peer-reviewed AC literature, there appears to be a significant AC modulation in which the average effect size (ES) when the experiments occurred at about 13 hours LST is around 4 times the overall average ES, while for trials at about 18 hours LST the average ES is zero. If this variation is not attributable to faulty statistical analysis, or to a systematic error produced by the non-uniform distribution of trials in time of day or day-of-year and is replicated in future data, it seems to me of some interest. We know very little at all about what physical variables in the environment correlate with psi performance. This observation may therefore be of some use in the future in building a successful model of the mechanism of psi. (Personal communication, 2003)

It is not yet clear whether this result has withstood the test of time and additional research. Certainly the peak at 13.50 hours LST showed up in those two separate batches of data, and its deviation from the background was statistically significant, but we must note that the number of trials involved was quite small. It's conceivable that the LST effect is real but strangely intermittent; it's also possible that it was just a random fluctuation. The more diverse oddities one looks for in a given batch of data, the more likely it is that something will pop up, like a face in clouds. That's okay, and science is the best system yet devised to determine whether such oddities are real or accidental.

Crudely, you gather another equivalent but independent batch of data and see whether the same anomaly jumps out. If it doesn't, this is good reason to suppose that the "effect" was just illusory.

Work by Dr. Bierman and a colleague in Amsterdam, Eva Lobach, saw the effect, although a reexamination of some Zener card work from the 1940s did not. Lobach and Bierman were attempting to replicate earlier findings by British biologist Dr. Rupert Sheldrake, who claims evidence that people really do have a sense of being stared at, and surprisingly often can sense who is calling them on the phone before they answer it—an awareness that seems to have cropped up as well with e-mails and text messaging. The Dutch researchers studied six women who were convinced that they had this power. Each completed six sessions of six trials, half of these inside what was hoped would prove to be the maximally effective LST window (the "peak" time), which was then between 8:00 and 9:00 in the morning in Amsterdam, and the other half at the non-peak time of 5:30 to 6:30 in the afternoon. They did not choose what the LST model predicted to be the worst possible time as their contrast with peak results, because that would have coincided with lunchtime, more difficult to arrange for participants.

Each participant nominated four relatives or close friends who agreed to take part, and at the selected times sat by the phone with one of the experimenters (who made sure there was no hanky panky), while the second experimenter randomly called one of the four possibles and arranged for that person to place a call to the subject within five minutes. Given that there were four possible callers, obviously the average success rate of randomly guessing who's ringing is 25 percent. In fact, the scoring rate at peak times soared to 34.6 percent, comprising almost all the anomalous results of the experiment, which achieved an overall success rate of 29.4 percent. That might have happened by chance one time in twenty. Once again, the number of events was very small, but the apparent sidereal time effect was quite marked.

Of course, given that the mysterious LST source is likely to be located in deep space, there is no way of knowing whether, akin to sunspot cycles, it fluctuates over the decades. If the peak effect is real, what could possibly be causing it? In a recent science fiction novel, I advanced a playful speculation that Spottiswoode disliked: I imagined a source of energy flux from the vicinity of the galactic center that has the effect of muting or even blocking psi.

> Start here: The human brain evolved in a fluctuating noisy environment. Most of the daily, monthly, seasonal rhythms, long ignored by medicine and psychology, are driven by solar cycles. But in the background of its functions, the human nervous system might be modulated as well by large-scale, feeble fluxes from beyond the solar system. As the [Sagittarius source] sets, the world's turning mass blocks its . . . radiation. Then a slow physiologic recovery peaks at optimum just around the next LST dawn.

Suppose that this worked (the novel suggested) by *suppressing* activities in the brain and nervous system relevant to psi, especially during the hours when these conjectural emissions are biologically detectable, unscreened by the earth's bulk.

> This inhibition would strengthen for several hours after sidereal dawn, peaking four or five hours later, as the nervous system clawed its way to equilibrium. Maybe suppression fell more slowly than it rose, reaching its lowest ebb again by the time the [galactic center] next lifted over the horizon. During this brief LST dawn-hour window, miracles worked best. . . . Then biological inhibition started to rise once again. . . . Worse still, perhaps the initial burst of . . . recovery skewed probability space sufficiently that when

[galactic center] hung directly overhead, the probability deformation relaxed into a brief countervailing rebound, then once again settled to near-chance levels as the Sagittarius radiator dropped toward and then below the horizon. No wonder psychics had such a hard time replicating their paranormal feats. (*Godplayers*, 2005)

But Spottiswoode told me: "There is *no* good fit to any model to a celestial source. At least that was the conclusion of the extensive mathematical modeling I did. I did most of my work to see whether distributed sources at a density equal to the distribution of emission from the galactic plane [GP] could be made to fit to the data on the model that the psi effect's strength was inversely proportional to integrated flux from the GP. The fit was not good. Single sources don't look good either" (personal communication, June 13, 2005). Currently, Spottiswoode is preparing a paper with Dr. Peter Sturrock at Stanford University using a new mathematical test that carries this investigation forward. All one can say at the moment is that the LST possibility is still in play.

❧

A rather different attempt to capture a possible global effect related to consciousness was devised by PEAR psychologist Dr. Roger D. Nelson, who retired from Princeton in 2002 but continued his innovative quest for signs of psi as manifested (he proposed) by the Jungian collective unconscious. He and his colleagues on the Global Consciousness Project (GCP) figured that if psychokinesis can sometimes modify the otherwise random output of machines built specifically to generate endless chance sequences of numbers, maybe it was possible that a bunch of scattered random event generators (REGs) left to their own devices might fall into synchrony, at least to some extent, chattering out their average tallies, their chunks of pluses and minuses, in matching patterns.

By the standards of conventional science it was a horrendous suggestion, because REGs are specifically designed to be isolated from one another, uncorrelated, the very definition of randomness. But suppose the minds of 6 billion humans and who-knows-how-many animals create a kind of shared tidal wash, ebbing and flowing with the daylight and seasons (making allowance, of course, for the geographical dispersion of these minds), but these days brought together at moments of crisis, shock and awe that are mediated by television, rumor, and the Internet. You can imagine that the death of a globally beloved figure might send a pulse of distress and grief through a substantial proportion of the world's population, with perhaps this shockwave racing slowly around the globe as the peoples of one nation fell into exhausted sleep and those in another awoke to the news of horror. Suppose there really is a collective mind of Gaia, comprised of us all. Might some tiny psychic capacity in even a small proportion of that immense number shake the random numbers coming out of innocent and mindless REGs as they churn away on benches in twenty or fifty labs or homes across the planet, registering coherence and resonance? It's a crazy idea, but not necessarily crazier than psi itself.

Nelson's project began slowly in August 1998, gathering steam, developing ever more sophisticated techniques for downloading, organizing, and analyzing data from some sixty-five host sites distributed unevenly around the world. Early results suggested to Nelson that some events of global note had indeed left tracks in more than one of these independent, true random event generators (*truly random* in the sense that these machines were not outputting streams of pseudorandom numbers that merely *look* random to the human mind but actually are completely determined by the single seed number with which they begin their calculations). In these gadgets, which came to be known fondly as EGGs (electroGaiagraphs, as it were), two hundred bits are pulsed out every second, an immense cumulative stream of pluses and minuses, ones and zeros. As with all such experiments, the default *non*-psi assumption, or null, is that over sufficient time an equal number of ones and zeros will be generated, with a certain variance that can be calculated from probability theory. Nelson's guess was that this variance might be shifted one way or another by collective perturbations in Gaia's consciousness, which is mostly comprised of human emotional responses. There was no reason to expect an excess of ones, say, rather than zeros, but you could look for excursions away from routine randomness.

The statistical details can be omitted (this sort of thing: "The focus for most analyses will be anomalous shifts of the segment mean. . . . The standard test for deviations from expected variation will be a Chi-square comparison of the composite deviation across all EGG's during the specified event against chance expectation. This composite will be a sum of the squared Z-scores for all EGG's and all predefined segments"). They are given abundantly on the GCP Web site, along with handsome charts and a running display of the output of the world's EGGs in real time. Apparently it doesn't matter to Nelson whether variance is greater or less than usual, just that there are "anomalous shifts of the mean during periods of time specified in formal predictions," some kind of nonrandom "structure." I have to say that strikes me as hard to justify, much less defensible than the ESP/PK proposition of psi-missing. I can't see any justification for thinking that on one day a global upset would cause way *more* noise than expected, but an equally upsetting event two weeks later would cause way *less* noise than expected. Unless, say, happy events calmed the fluctuations down while horrid events caused an excess of variance—which is not the case. Putting this objection aside, a key point that GCP looks for is simplicity itself: *Are there unusual variance spikes associated with topics of particular significance to entire communities?* These need not be all horrors, since holy days of prayer and meditation or wildly exciting sporting events might also be expected to nudge the EGG output if this crazy idea has any weight. What's more, is it possible that some precognitive foreknowledge of these globally significant events might show up as advance surges in the charts?

Nelson's answer is a cautious yes. There was a certain horrified excitement when analysis seemed to show an enormous spike corresponding with the American tragedy of the World Trade Center murders on September 11, 2001. Certainly if an enormity of this magnitude did not impress itself on the variance of the world's array of EGGs, one might conclude that the experiment had failed, and failed conspicuously. On the other hand, finding such a spike might not be conclusive evidence in favor of collective PK, of a "disturbance in the Force"—it might just be what happens when motivated people with bees in their bonnets trawl through a huge body of data fishing for whatever they can find. That, certainly, was the explanation offered by critics, who regarded the mathematics of the GCP with scorn and, of course, not a little hostility. More interestingly, the former SRI/SAIC specialists in anomalies, hardly foes of the psi hypothesis, rejected the GCP analysis uncompromisingly. May and Spottiswoode demonstrated that the shockingly vivid upward gash on the paper at around the time of the attacks on the World Trade Center depended upon the data being analyzed in one particular, rather arbitrary way—in regions comprising four hours. When the analysis is based on, say, three-hour periods, the test hypothesis fails to achieve significance. "Although there is a single [highly significant] 1-second . . . in the middle of the New York attacks, we find that it is completely consistent with chance expectation and the distribution of z-scores. Furthermore, it has never been the claim that the EGG network would 'feel our pain' for just a second and move on . . . if our analyses and interpretations of the data are correct, then it is our view that the worldwide network of EGG's did not respond to the terrible events of September 11, 2001."

What's more, other drastic and startling peaks can be found throughout a long record of the EGG output, but few of those correspond to globally meaningful events. On the other hand, it's worth noting that GCP claimed a significant spike in the vicinity of the funeral of Diana, former wife of Prince Charles. ("During the public ceremonies for Princess Diana, the data taken in 12 independent recordings at various locations in Europe and the United States compounded to a significant result indicating an anomalous global effect which would occur by chance only about once in 100 repetitions of an experiment of this nature.") Granted, that event spammed Western newspapers and airwaves, and certainly was of powerful emotional significance in Britain and to a lesser extent in the United States—but might hardly be thought to have much impact in, say, China, where a very large proportion of the earth population lives. Then again, maybe it's sufficient to have many millions of people simultaneously if locally brought to tears of grief or happiness in an overwhelming moment of selfless empathy, wherever they happen to be in the world, whether or not other cultures share their pain or joy. The best one can say about this ongoing project is probably that it is an intriguing idea, consonant with the notion of psi as a widespread human facility linked especially to our emotional responses, our shared human empathy, our sense of danger of group exhilaration. Major international sporting events arousing turbulent emotionality are also under GCP's spotlight, but, oddly enough, World Cup Soccer and other major, globally monitored sporting events "have generally not produced persuasive effects on the EGG network." Dr. Radin has pointed out to me that sports games do not promote collective mental coherence, which is what the GCP might be tracking. Rather, sporting games deliberately provoke conflict, which presumably fails to promote the right type of attention.

☙❧

Moving radically from the planetary to the everyday, from the cosmo-
logical to the physiological, psi researchers have been looking to see
what kinds of events in normal life might give rise to bursts of paranormal knowledge or action. Recall the anecdote with which I opened
this book: my wife, Barbara, subjected to a shocking and emotionally
charged vision of a bloody body, correlating unexpectedly with the
fatal car accident of her cousin Lizzie, a woman who was in effect her
twin sister. If psi is an evolved function, we'd expect it to alert us to
danger, and perhaps prepare us for terrible events that can't be
changed. Supposing that this is so, it seems plausible that something
like the mechanism used in lie detectors might detect small changes in
our body's autonomic preparedness, in the ancient templates of
arousal that ready us for flight or fight. Such changes in the body can
be drastic—the wide, pinpoint eyes, blood draining from the skin to
the safety of the inner organs, bursts of adrenalin and cortisol, hair
bristling, muscles charged for action—or much more muted, yet even
so observable, especially when subtle instruments are placed upon
our skin or scalp. Better yet, a functional magnetic resonance image of
instantaneous slices through our brain activity—but that is expensive, very expensive, and beyond the reach of most parapsychologists.

Cheap computers have made all the difference. Now you can
track the rise and fall of whatever index you choose to investigate,
such as sweating, which alters the electrical conductivity of the skin
and correlates significantly with levels of stress, and then run the history of these traces as a moving graph. Off-the-shelf programs can
take this data and sort it in many different ways, smoothing out
random surges in current or resistance, removing noise from the
signal. It seems too simple, and yet perhaps it's just simple enough,
now that all the complexity is hidden away under the hood of the
desktop computer. This approach has given rise to a search for a phenomenon dubbed *presentiment:* bursts of autonomic activity that are
pretty much outside the control of ego consciousness, but which

anticipate small shocks to body and brain, literally anticipating by a second or two, rather than registering those shocks after the stimulus happened.

The idea could hardly be more basic. Wire up a nervous friend, put her at ease, then betray the trust by having her walk into a dark room, shout "Boo!" Of course, your friend's heartbeat accelerates, she jumps, she cries out in startlement. All this is commonplace, a rather crude, cruel practical joke. But suppose when you look at the tape of her reaction, you find that her body has gone to alert not half a second *after* you yelled in her ear, but a second and a half *earlier*. Does this ever happen? Even if it did, it would be hard to be sure that your luckless friend had not detected some noise in a darkened room, smelled your presence, heard your breathing. The same applies, even more so, if there's a hungry waiting tiger you placed in a cage in the room. So if we are looking for evidence of psi presentiment, as usual we need to strip away almost all of the very circumstances that would make psi useful, even necessary, in the real world. Especially if the real world held ravenous animals up every tree, which is a reasonable approximation to the kind of life our evolutionary ancestors adapted to for a million years.

Since you are using a computer to analyze the data, you might as well use it also to provide the shocking stimuli that come bursting out amid the soothing background that is meant to loll your subject into a false sense of security. That's just what parapsychologists such as Dr. Dean Radin and Dr. Dick Bierman have been doing since the mid-1990s, following on earlier experiments that had been conducted intermittently during the previous twenty years. Radin used one of the simplest possible methods, measuring shifts in skin conductivity in fingers or palms when the wired subjects were affronted by computer images of violent or erotic scenes, and those reactions in turn were compared with the physical response elicited by both soothing images and no images at all. When the data from many subjects were added together and averaged, in order to remove idiosyncratic responses and the intrusion of random noise, it turned out that the average response to neutral or pleasant images followed pretty much the curve one would expect. The image flashes on the screen for three seconds, and while subjects watch the blank screen that follows, their skin conductivity rises slightly, drops away again, and then flutters along in its normal quietly wandering path. After emotionally charged topics, though, skin conductance soars to a quick peak moments after the image has flashed up, then again ebbs away as the subject recovers from the brief startle or shock. All of this is only to be expected by any physiologist. Radin's and Bierman's remarkable claim, though, is that the emotional images appear to cause a smaller anticipatory surge *before* they are displayed *(presponses)*—in some cases even before the computer has *chosen* them from a random pool. It's precognition on a small scale, registered by tiny currents that participants can't even feel.

This paradigm was eventually extended from simple lie-detection devices that look for modulations in galvanic skin response to the more complex brain-scanning devices used by medical physiologists, brain surgeons, and cognitive scientists in their quest to understand how the various parts of the brain function together. The great thing about this approach is that a huge trove of data already exists, precisely the research materials of scientists looking for almost anything except psi. When Radin and his colleagues accessed this material, for the most part their findings had been replicated in advance. Rather suitably, Bierman reexamined old studies on phobias and gambling behaviors and found small but significant prestimulus rises in the ways people reacted to, for example, calm images versus pictures of animals or erotic scenes (even among the phobic, the naughty pictures cause more of a leap than the scary animal shots, probably something Darwin would have predicted). An excellent description of such presentiment research can be found in Dean Radin's book *Entangled*

Minds, where he quotes Nobel Prize winner Kary Mullis, who visited his lab in 1999: "It's spooky. You sit there and watch this little trace, and about three seconds, on average, before the picture comes on, you have a little response in your skin conductivity which is in the same direction that a large response occurs after you see the picture. . . . That, with me, is on the edge of physics itself, with time."

After a dry run on his own brain, Bierman went more high-tech, using a noninvasive instrument called Blood Oxygenation Level Dependent fMRI. This provides pretty color-coded pictures of blood oxygen levels in the brain as a subject responds to certain stimuli or performs a simple task. Bierman chose the by-now-standard tripolar workhorse of three kinds of visual stimulus—calming, violent, erotic—drawn from an equally standard image inventory. He was flashed a sequence of images for 4.2 seconds each, from a selection of 18 violent, 18 erotic, and 48 calming images. Oddly enough, there was no presentiment elevation before either the calm or the violent pictures, but the lift created by the erotic pictures was improbable by chance at the level of some 1 in 320, certainly significant.

Encouraged, he applied the test to six male and four female volunteers, segregating their results according to sex. The average male reaction resembled his own. Again, there was no special arousal in advance of violent images, but a barely significant response to the erotic pictures. The females did react to the erotic stimuli, but even more strongly to the violent ones. What this tells us about our cultural conditioning and our inherited propensities might be worth musing upon. Given the very small number of subjects, it is remarkable that Bierman got any kind of significance at all from his results, but in fact the combined erotic target results were improbable at the level of 1 in 250. Unfortunately, the fMRI portraits of these anticipating brains showed no lit-up activation of any specialized "psi lobe" or "neural psi module." Psi, apparently, like many other high-level functions, is distributed across the neurological structures of the brain.

⊚⊚

In the conductance experiments described above, the participants selected their own starting times for each image display. Some years later, Ed May and James Spottiswoode developed another method for testing presentiment (which they renamed *prestimulus response* or PSR) (*JSE*, 2003). Their participants sat quietly in a comfortable chair wearing a sound-isolating headset providing a low-level acoustic background. Every now and then a blast of ninety-seven-decibel white noise would crash randomly into the earphones for a second. Alternatively, they might hear one second of complete silence. As previously, skin conductance measures were used to represent the amount of arousal created by the noisy blasts. The presentiment element to the experience was an expectation that the collective or average resistance—the "non-specific skin conductance response"— would rise in advance of the noise to a greater extent than it did before the silent periods.

It was an intensely simple and elegant experiment. The result, as predicted, agreed with the presentiment hypothesis. The heightened curve of anticipatory response before the noisy signal differed sufficiently from the very muted response before the period of silence that the difference could be expected by chance, according to their calculation of probabilities, only once in more than a thousand repetitions of such experiments with 125 participants. (Their statistics were challenged by Mikel Aickin in an accompanying commentary; interestingly, Aickin did not reach a different conclusion, but regretted certain formal lapses. The most interesting suggestion Aickin advanced was that the results might not be due to precognitive prestimulus awareness but rather to a different psychic effect, real-time PK influence over the random number generators, possibly by the experimenters themselves.) These prestimulus response findings—if that's what they are, if they are not, after all, due to psychokinesis—

have since been replicated significantly, if just barely, with fifty participants by May and two Hungarian colleagues, Tamás Paulinyi and Zoltán Vassy.[1] More recently still, May has refined his setup and is currently looking for similar effects in variation of heart rate.

Meanwhile, May and his colleagues had gone looking for ways to beef up psi accuracy, deriving their hunches from information theory. For a start, any AC protocol that interrogates a complex scene so coarsely that only a binary response is acceptable (rainy or fine? night or day? male or female?) inevitably throws away a considerable amount of information. The development of fuzzy set theory by Dr. Lotfi Zadeh in the 1960s and '70s provided a formal way to distinguish a scene that is, say, *predominantly* shady from one that is predominantly bright, and those in turn from scenes that are *considerably* shady or *glaringly* bright. You can see why this sort of fine-grain discrimination would be useful in psychic spying. It's also beneficial when trying to establish the working parameters of psi.

There are other measures of any possible target scene, such as how visually or conceptually busy it is, how much variation can be detected between one part of the scene and another, and so on. May and his associates homed in on a measure known as *Shannon entropy* (named for the great telecommunications scientist Claude Shannon, who created much of information theory more than half a century ago)—in particular, estimates of entropy gradient. Entropy measures the amount of uncertainty, or lack of precise information, an external observer has about a system. In fact, in formal terms (the equations developed by Shannon), information is identified with *negative* entropy. If entropy measures the variety of possible states of a system, then any single state chosen from a system of high entropy must be highly improbable. If we observers can specify it precisely (identify it, in short), we thereby massively reduce our uncertainty about the state of the system. Hence, the information conveyed by this exact specification will be correspondingly great. By contrast, if a system has only two possible equally likely states, specifying which of those states it is in will reduce uncertainty only by one bit—that is, the amount of information conveyed by a single binary choice: yes or no, on or off. The corresponding reduction in entropy is minimal. Of course, because this schematization explicitly avoids the *meaning* that a message recipient might ascribe to the several states under consideration, this sense of "information" is somewhat paradoxical. Clearly, it is humanly more significant to know if one will live or die when facing the barrels of a firing squad (a minimal reduction in uncertainty) than to discover which of ten million virtually indistinguishable flies is closest to the top right-hand corner of a box. Information theory is too simple to account for this; it is a theory of telephone line efficiency, not of the vibrant people who telephone one another.

Anyway, May and his colleagues found a method for estimating the quantity of entropy in one part of an image or a scene and comparing it with equivalent quantities elsewhere in the scene. The more dynamic or surprising the target scene—the larger the degree of variation in intensity of its components—the greater the formal entropy gradient turns out to be. Since this evaluation can be conducted dispassionately by a computer program, in advance of human inspection, the way was open to selection of optimal targets with maximal entropy gradients. Testing this approach against existing databases of remote viewing and other psi scores confirmed the theoretical prediction that scenes with high intrinsic entropy gradients made better targets—that is, they were more often associated with correct calls or descriptions.[2]

☙❧

If I have tended to emphasize results obtained by those involved for many years with the secret American government psi projects, that's partly because of their impressive results and also because many of us share a background assumption that if you throw lots of money at an obstinate problem, you'll probably have a better chance of solving it. One depressing counterexample that Ed May mentioned to me is the expenditure of some half million dollars in an attempt to create an environment for testing psychokinesis under absolutely stringent conditions. The first thing researchers discovered while building their elaborate test equipment was that you can't exclude all possible variables. The second was that no psychokinesis manifested itself under such highly scrutinized conditions. Those who remain skeptical of the very existence of psi will not be surprised. Obviously, they point out, as you increase the precautions against accident or fraud, the likelihood of seeing apparently anomalous effects must diminish sharply. This claim has been rebutted repeatedly by parapsychologists. One bold attempt to circumvent the whole issue is a rather charming experimental protocol developed and tested recently by Dr. Suitbert Ertel, professor emeritus at Georg-Elias-Müller-Institut für Psychologie in Göttingen, Germany.

Ertel looked for a method that would sidestep the tedium of traditional Rhine-style card-guessing tests, while retaining the underlying statistical merits of a repeated task. The best way to do that, he thought, was to make it fun. What's more, sitting in a chair and guessing thousands upon thousands of times at a choice of shuffled cards is not the sort of thing the human mind and the human body have evolved to excel at. We like to scan the world for what interests us, and then reach out and grab, strike, or caress. That's why the computer mouse works so well as a gadget for controlling the position of a cursor as it darts across software displays, compared to some abstract method of identifying a position on screen by specifying, say, numbered coordinates. Dr. William Calvin has argued that the human mind developed in conjunction with the human hand and arm, taking up stones and hurling them with increasing agility and accuracy at games and, inevitably and lethally, at one another. This linkage of physical prowess with mental capacity is undeniable. It's the underlying mechanism that makes computer games so enthralling and satisfying. You can watch a movie and be battered with insanely expensive special effects, and that has its thrill; you can sit quietly and read a novel, dreaming your way through a kind of emulation of the events in the story, and that's very satisfying as well, to those who have learned how to do it. But the development of interactive computer games is rapidly overwhelming both of these now-traditional pastimes. Kids sit before their Xbox or PC monitor, fingers gripping the tactile joystick or controller, clicking like lunatics, immersed physically as well as mentally in a fast-moving simulation of reality that seems realer than real. So Suitbert Ertel looked for a simple, mildly amusing task that conscripted this ancient hand-eye coordination. While it lacks the visual pizzazz of *World of Warcraft,* it certainly seems to captivate the student participants. (Surely it won't be long before complex computer-game design is adapted to this kind of psi task—at which point I expect the results to become truly dazzling.)

Ertel hands you an opaque bag. Inside are fifty ordinary table tennis or Ping-Pong balls, with numbers written on them—ten each, numbered one through five. Your task could not be simpler: announce, and note down in advance, your guess at the number marked on the next ball to be drawn, shake the bag, reach in and rummage about, seize one ball that feels right to you, and pull it out. Take note of the number and record it in the column next to your guess (or, of course, have somebody else take these notes; better yet, have the whole thing conducted under videotape). Replace the ball, make another guess, shake the bag, take out another numbered ball, rinse, lather, and repeat for sixty trials. Put so drably it sounds about as much fun as counting your fingers and toes for an hour and a half. Apparently the task is not nearly as dire as one might fear. An element of challenge arises, since participants are getting immediate feedback and can tell how well or poorly they're doing.

Remarkably, this method produces robust and highly significant results (Dr. Ertel assures us). In a series of experiments involving five groups of first-year psychology students, 238 in all, with a mean chance expectation of getting just 20 percent correct, 1 in 5 of the students was successful at the 0.05 level (by definition, there should have been only about 1 in 20 succeeding that well), and 1 in 10 at the 0.01 level (about ten times more than you'd expect by chance). But critics broke out in hives when they read that in the first pass Ertel was allowing his students to take their bags of balls home, where they ran the experiment alone, guessing and recording the results without an overseer. This looks like a flagrant prescription for rampant cheating and fraud. Ertel acknowledged that these data were not safe, so he also insisted upon a second phase that served as a control on consistency: rather than a precaution, a *post*caution. Those students who scored remarkably well were invited back into the laboratory for further tests under controlled conditions. You would expect two things to happen: even if most students were completely honest, some would have done unusually well in their home testing purely by chance or as a result of inadvertent recording errors, so their scores the second time around would probably revert to chance. And those who did cheat would also revert to chance scoring, unless they were especially adept at subterfuge. If there is no such thing as psi, that would exhaust the possibilities (except for an extremely small number who by chance did well the second time). That's not what happened.

Even to the experimenter's surprise, hit rates for the high scorers at home remained highly significant under laboratory conditions— "six of ten high hitters under home conditions had significant hit surpluses under lab conditions"—with a chance probability of less than 10^{-15}. That is, one chance of scoring this well in a thousand million million repetitions of the experiment. Overall, these novice students scored 8.8 percent better than the chance expectation. And the speed with which the best of them attained the standard significance level of 0.05 percent was markedly greater than parapsychologists expect using other formats: "Michael . . . who has the last position among the best scorers, reached the benchmark in 41 minutes. Claudia on top of the rank order . . . made it in 2 minutes" ("The Ball Drawing Test. Psi on Untrodden Ground." Based on a paper presented at the 43rd Annual Convention of the Parapsychological Association in Freiburg, Germany, August 17–20, 2000).

Dr. Ertel notes that if this were to be done by fraud, you might expect the ball that was just drawn to be held "craftily in one's hand without putting it back," allowing the cheat "to call that ball's number on the next trial. But the number of hits for just-drawn numbers . . . is not larger than the number of hits for not-just-drawn numbers" (Ertel, 2005). Might sensitive fingers detect the shape or color of the numbers written on the balls? Perhaps, but wearing thin cotton gloves does not obliterate the effect, although gloves can diminish it—perhaps, Ertel conjectures, because participants feel separated from the balls as natural contact with the balls is obstructed.

What's more, "The ball test material has been examined by German skeptics (GWUP) whom I asked whether they could tell me how participants might obtain hits by deception. They were suspicious, but, despite my insisting to tell me by which tricks a person could obtain hits they lacked an explanation." Variations on this basic ball-drawing experiment have been tried out, seeking insight into the processes involved. One approach has been to use five unnumbered beads of different colors; another had added red or green dots to the numbered Ping-Pong balls, with the task being to identify both the number and the color (Ertel, "Psi Test Feats Achieved Alone at Home: Do They Disappear Under Lab Control?" 2004). It seems that for targets displaying both numbers and colors, the amount of effective psi gets split between the two elements, consistent with the kind of differential effect long recognized in parapsychology.

As usual, and as is proper, it is people from within the anomalies community who have been most critical, even suspicious, of Ertel's results. To my mind, the criticisms to date have not been damaging. And with results as good as this—"for three high scorers at home, for Ahmed, Amelie, and Silke, hit rates under control conditions even increased immensely"—and from a test so simple and easily replicated, you might wonder whether Suitbert Ertel has come up with a challenge that even scientific skeptic James "The Amazing" Randi might quail at. The conjurer and his team offer $1 million to anyone who can demonstrate paranormal effects to Randi's satisfaction, but they insist upon a gigantic deviation from chance expectation.[3] Not just one in one hundred or one thousand, but one in a million—and not just once, but several times. Worse still, Ertel reports, they have insisted on changing or actually abolishing test conditions he regards as psi-conducive.

If Ertel's protocol remains robust and repeatable, he and his high-scoring amateur psychics might be able to take the money away from the amazed skeptics.

✑

There are reasons, alas, to doubt that any given protocol will in fact remain robust and repeatable. This is surely one of the main reasons psi remains barricaded on the wrong side of the gates of science. Skeptics claim that replication problems are due to early apparently positive results being nothing more than errors, fraud, or simple coincidence, swiftly erased when the errors are located and corrected, the frauds hounded out of the laboratory, and coincidences dry up, as they must in any random process.

Plenty of close analysis by parapsychologists has demonstrated the futility of these simple explanations. Patterns were found early by such researchers as Dr. J. B. Rhine, who noticed in the handwritten records of card-guessing experiments that results tended to be abnormally positive at the start of a run, fall away toward chance, or even below, but then rebound somewhat, forming a sort of *U* shape. Rhine recommended attention to this modified-decline effect as a signal of the activity of psi, and indeed it was found subsequently in earlier records when they were examined. These days, Rhine's ready acceptance of this curious sag and rebound is cast into doubt by the discovery (or at least the claim) that later observers can exert a psychokinetic influence backward in time, modifying or corrupting whatever pressure the participant was attempting to exert. If that kind of temporal entanglement turns out to be important, it's possible that Rhine and his obedient followers were themselves creating the U-shaped pattern in the old results. This way lies madness, you might suppose. Such "explanations" have the distinct look of cop-outs, post facto rationalizations. But decline effects are ubiquitous in anomalies research. Attention must be paid.

In 2001 Dr. Dick Bierman published a most unnerving study of failures to replicate apparently well-established anomalous phenomena. What he seems to have demonstrated is not just that psi declines within a given experimental run, nor even within the history of a particular participant or experimenter (which might be attributable to boredom or exhaustion), but within whole paradigms of research. Mapping six different paradigms—attempts to exert mental control over tumbling dice, early use of the ganzfeld telepathy protocol, subsequent improved protocols using completely automated elements, precognitive card guessing, attempts to manipulate random number generator output by directed attention, and attempts to manipulate living systems by directed attention—Bierman showed vividly how over time the scatter of experimental scoring created a downward curve toward the mean and nonsignificance, and usually then continued on below the mean (negative scoring). It was as if the light was dimming from one decade to the next, eyes struggling to compensate, hands reaching uneasily, knocking things over, blindness. And all of this working unconsciously, for most of us.

With the dice throwing, results hit the mean around 1979 and dropped under it. Early ganzfeld tests slipped down to the mean a little after 1988. More recent and sophisticated ganzfeld experiments dipped savagely in the mid-1990s. Something similar clearly occurred in attempts to modify the responses of living creatures, although precise dates are not available. Could this be some kind of secular oscillation or cycle, akin to the LST or to solar cycles marked by sunspots (but longer), or the great cycles of global heating and cooling (but very much shorter)? If so, one might expect to see a rebound from these dismal declines, should they continue long enough. And indeed, that is pretty much what Bierman found in the majority of databases recording results of attempts to modify random number generators by PK. Here the slow attenuation from 1955 to around the middle of the 1970s slowly reversed and climbed again to previous levels of effectiveness by 1990. In the center of that curve was a decade where average results were flat, although, of course, individual experiments continued to manifest scores above and below the mean.

More careful examination of several of these databases also showed a certain extended U-shaped pattern of decline and recovery.[4] But the structure of experiments are also importantly variable. For example, Roger Nelson has commented:

PEAR REG work does have declines (and some inclines, to be sure). But also the experiment evolved through several major versions, and there are a number of changing aspects even in the most basic manifestation: The Operators (participants) were numerous, and our protocol allowed them to take most of the 'experimenter' role; several potential modulators were always available as options (source of instruction, number of trials per button press, nature of feedback, etc.); remote and offtime operation was tested; and so on. Also, and most important, the attitude and intention of the PEAR REG program was perhaps a bit different from what you might find in some replication efforts. For me it was nicely expressed in a Sufi saying: 'Have fun, or try to learn something. If you do, you will annoy someone; if you don't, you will annoy someone.' (Private communication, February 11, 2007)

If long-term decline and recovery trends are corroborated, we'll have another quite striking psi regularity, one not observed or expected from random coincidence. What could be responsible? That is a question requiring a theoretical answer, one that will drive experiments in the direction of greater specificity, accuracy, and, well, with luck, some satisfying explanations for this damned irritating phenomenon.

5

THEORIES OF PSI

The objective of the present work is to present a complete theoretical foundation or paradigm, consistent with, indeed springing from, physical laws and principles that will encompass paranormal phenomena.

—Dr. Evan Harris Walker

Work in this field is a complete waste of time. Although it is politically incorrect to dismiss ideas out of hand, in this case there is absolutely no reason to suppose that telepathy is anything more than a charlatan's fantasy.

—Dr. Peter Atkins

In September 2006 the science editor of the *Times* of London reported shocked uproar created by a public session favorable to parapsychology, run under the auspices of the British Association for the Advancement of Science (the BA) during that nation's "premier science festival." The BA festival's organizers "were accused of lending credibility to maverick theories on the paranormal by allowing the highly controversial research to be aired unchallenged."[1] Interestingly, the outraged critics cited were all given their titles (Dr., Professor, even Lord), while those on the session panel, including Dr. Rupert Sheldrake (a former research fellow of the lofty Royal Society) and Professor Deborah Delanoy of the University of Northampton, went without.

In light of the evidence we have considered so far, the objections quoted seem no better than tantrums and self-confessed ignorance. A sidebar quotes fertility specialist Lord Winston, a former president of the BA: "'I know of no serious, properly done studies which make me feel that this is anything other than nonsense. It is perfectly reasonable to have a session like this, but it should be robustly challenged by scientists who work in accredited psychological fields.'"[2] A delicious exchange in a subsequent BBC interview with chemist and science writer Dr. Peter Atkins and Dr. Sheldrake makes the point even more clearly:

Interviewer: On the other hand when [Sheldrake] produces his evidence, he said 25% was what you would expect, but what he got was 45%, that is remarkable.
Atkins: No, that's just playing with statistics.
Interviewer: Let's put that to Rupert. Rupert Sheldrake, he says you're just playing with statistics. He doesn't believe a word of it. What do you say to him?
Rupert: Well I'd like to ask him if he's actually read the evidence? May I ask you Professor Atkins if you've actually studied any of this evidence or any other evidence?
Atkins: No, but I would be very suspicious of it.[3]

Although participants on the panel noticed no furor, the fuss is reminiscent of what happened in 2001 when the Royal Mail in Britain published a special brochure to accompany their issue of special stamps to commemorate British Nobel Prize winners. Dr. Brian Josephson, Nobel physics laureate in 1973, took the opportunity to draw attention to anomalies research: "Quantum theory is now being fruitfully combined with theories of information and computation. These developments may lead to an explanation of processes still not understood within conventional science such as telepathy, an area where Britain is at the forefront of research."[4] *Nature* was more amused than affronted: "But few physicists accept that telepathy even exists, says Andrew Steane, a quantum physicist at the University of Oxford. Robert Evans, a physicist at the University of Bristol, says he is 'very uneasy' about something from the Royal Mail saying quantum physics has something to do with telepathy."[5] Josephson responded in the *Observer* newspaper on October 7, 2001:

> The problem is that scientists critical of this research do not give their normal careful attention to the scientific literature on the paranormal: it is much easier instead to accept official views or views of biased skeptics. . . . Obviously the critics are unaware that in a paper published in 1989 in a refereed physics journal, Fotini Pallikari and I demonstrated a way in which a particular version of quantum theory could get round the usual restrictions against the exploitation of the telepathy-like connections in a quantum system. Another physicist discovered the same principle independently; so far no one has pointed out any flaws.[6]

An academic and science correspondent for the London *Sunday Telegraph*, Robert Matthews, commented sharply in November 2001: "Just consider: there is no credible evidence that time travel has ever been achieved, but that has not stopped serious scientists pondering ways in which it might be. In contrast, there is now a wealth of evidence for the existence of ESP, obtained by researchers from reputable universities on a repeatable basis. Yet, any scientists who dare suggest ways in which ESP might be possible can expect a heap of ordure to be tipped on their heads by fellow academics."

Really, the objection that conventional scientists raise against the idea of psi is not that the evidence is deficient. Most of them have never looked at it, carefully or at all, although many note acerbically that they see no sign of psi disrupting the results and meter readings in their own labs. More crucially, the motive for dismissing psi is that the reigning theories of science (or so it's asserted) do not leave any room for psychic phenomena. Albert Einstein, who at times expressed an interest in such anomalies and even wrote a preface to Upton Sinclair's *Mental Radio*, eventually rejected the topic as unscientific when he learned that ESP failed to obey the inverse square law (falling off sharply with distance), and therefore could not be regarded as a form of transmission akin to radio. That was a quite remarkably limited view of the possibilities available for scientific explanation, but then Einstein didn't like quantum theory, either. Indeed, in a January 2004 debate on telepathy sponsored by the Royal Society of Arts, Dr. Sheldrake observed:

> There's no inverse square law [in psi or the quantum theory of nonlocality]. When Einstein first realized this implication of quantum theory, he thought quantum theory must be wrong, because if it were right, it implied "a spooky action at a distance," as he put it. It turns out quantum theory is right, Einstein's wrong and that particles or systems that are in part of the same system, when apart, retain this nonlocal connection. . . . If quantum theory is truly fundamental, then we may be seeing something analogous, even homologous, at the level of organisms. Insofar as people are thinking theories of telepathy, then this is one of the prime contenders.[7]

In that debate, Sheldrake's opponent, anatomy professor Lewis Wolpert, offered the standard complaint, after first ritually denying that any acceptable evidence can be found: "I suppose, as a scientist, it's slightly weird that what the people [do] who work in this field is just to provide more examples. They make no effort whatsoever to understand what's going on." Although Sheldrake replied: "There's no shortage of theoretical work in this area," still the argument typically leveled against the reality of psi by scientists is that there's no sound theory to support it. Raw observations are not enough. You need a powerful and principled theory to constrain your observations, to predict in advance what will happen reliably under exact circumstances, and, just as importantly, what *can't* happen, and why your story is better than the other guy's.

It's true, therefore, that the study of psi anomalies could use a few good theories or, as they're sometimes known, robust research programs. It's not that nobody bothers to think about what they're doing in parapsychology research. Every experiment derives from a *hypothesis*. After all, any spontaneous anomalous experience implies a minimal hypothesis, at the very least a suspicion that everyday expectations have been cruelly violated: "Something is happening here, and I don't know what it is, the standard rules are being broken, something is afoot, what the hell is going on?" But hypotheses, even quite clever hypotheses, are a dime a dozen. Genuine *theories* are much more costly. They are capacious containers, filled with compartments that seem to subdivide endlessly as you get inside them. What you put into those compartments is every relevant observation you've ever made, and then you look to see what you can find in the world to fit into some of the empty compartments. If there is anything left over, you could be in trouble. And if your theory box holds many yawning compartments that look as if they will never be filled, you might be well advised to get hold of another container that uses a different design and start again.

The best available guide to psi theory remains a very large review paper from 1987, Douglas M. Stokes's "Theoretical Parapsychology." Inevitably, more recent discoveries (such as presentiment and LST) are missing, as are theories prompted by work developed under the

Star Gate program and elsewhere (to which we'll return). Still, this 110-page document bristles with information on the ways in which paranormal phenomena have been analyzed theoretically, if never very successfully. Here are the alternative categories Stokes uses in cataloguing these approaches: First, that psi doesn't exist except as delusion, sloppy errors, or outright fraud. Second, that it's real but requires modifications of current understandings of spacetime: backward causation, branching time, the literal reality of quantum theory's many worlds. Third, signals of a known—or more probably an unknown—kind might account for information or energy transfers. Fourth, theories derived from quantum theory explicitly, emphasizing nonlocality, entanglement, and even the crystallization of reality through observation. (This is the category Stokes found especially enticing: "A most exciting development in the past five years has been the experimental confirmation of the principle of nonlocality in quantum mechanics and the realization of the importance of that principle for a theory of psi phenomena.") Fifth, theories proposing that the mind is external to the material brain, and in coupling with it brings extra dimensions of freedom. Sixth, perhaps best seen as a subset of the previous models, neo-vitalist theories of a spirit that might survive physical death. More recent models, such as the decision augmentation theory (DAT) of May and his colleagues and other researchers that gesture at superstring or brane theories from high physics, can be assimilated into these general pigeonholes. None seems especially robust at this time, able to explain not just psi but how everything else known to science is neatly and testably consilient with its outrages.

☯

Before you can start developing a theory of psi, you need to think very carefully how the observed phenomena fit in with what we know from daily life and from existing scientific research. We see the world because photons of various frequencies are emitted from energetic light sources and then rebound from objects before skidding through the lenses that focus them on our retinas. That activates a cascade of electrochemical activity, including a lot of cunning preprocessing, before the reduced signal is passed along to specialized sections of the brain, themselves honed by millions of years of selection. Smelling a scent is superficially very different: a chemical is carried on the air, or perhaps the fingers, to the nostrils, where similar chains of electro-chemical activity ensue, traveling to quite different portions of the brain. Taste does something quite similar, but emphasizes different chemicals. Hearing involves compression and decompression of the air funneled in through ears, and sometimes bones, where again information is extracted and coded into nervous system data. Touch and balance and digestion and all the other senses (we have many more than the traditional five) perform this transmutation of impacts and chemical identifications, leading to a swarm of pulses into the brain. None of our senses is magical; each uses the micro-scopic equivalent of a physical jolt, translated into the neural coding that the long-evolved brain can recognize and interpret, given years of experience and training in infancy and childhood. So what's the equivalent for psi? Where are the input channels for the "extrasen-sory" senses? Where are the output structures enabling us to work our psychic will on the external world? I can easily hear you speak, and even understand you if we share a language, but I can't hear you *think*. How would that work?

But are we actually as isolated as this? In my 1980 science fiction novel, *The Dreaming,* I playfully suggested that while neuroscience assumes the brain is the seat of consciousness, perhaps memory and thought and will are not just functions of this biocomputer of 10^{17} bits of storage capacity. "What the theorists have neglected is the con-nection between human beings, the constant flux of data we transmit and receive in the ultralow frequency band," says one of the characters. "The human brain and nervous system . . . are an indexing system, a file of coding procedures, a folded aerial a million meters in length. The deep grammars of thought and language are a common property spread with massive redundancy through the brains of every human alive on the planet. We are one another." In my fiction, this constant flux of shared information is stored inside an immense neutronium crystal left on earth by visitors. It is a technological embodiment of what ancient Hindu philosophers called the "Akashic records." Even if some such technological marvel did exist, constituting a sort of colossal planetary Google, it seems unlikely that ultralow-frequency emissions from the brain and nervous system could cope with the urgent requirements of daily life. Still, perhaps some such central repository exists, generated somehow in a higher-dimensional space (whatever that means in the era of string theory and its curled-up dimensions) during the long evolution of humankind. If so, perhaps we all have access to it, however primitive our uninstructed approach may be. It is not as if we spring from the womb spouting Shakespeare and differential equations; we have inherited the raw capacity to learn these rich discourses, but that learning is inscribed in culture, not in our genes. It might be the same with psychic capacities, even if our very minds are partially located in a shared archive.

Does such a Matrix exist, or is it just a metaphor, like Swann's "signal line" into which his remote viewers were meant to tap? Joe McMoneagle tells us that the Matrix was always nothing more than a figure of speech. Oddly enough, PEAR's Robert Jahn and Brenda Dunne took the notion far more seriously. In their 2004 paper "Sensors, Filters, and the Source of Reality," they venture upon mysticism, proposing that "our palpable physical surround is an emergent property of a much vaster intangible reservoir of potential information, which we have labeled the Source, and that the emergence is enabled by the resonant coupling of this Source with its cosmic complement, the organizing Consciousness."

The Source. Oh dear. Well, why are we not normally aware of the Source? Like Aldous Huxley and the poet William Blake before him, it's a matter of clogged filters and dirtied doors of perception. "We have suggested that the intensity of that resonance is limited by the physiological and mental filters imposed upon our objective and subjective sensory channels by various physical, cultural, and emotional factors." Might we relax or tune those filters "to enhance the resonant dialogue between Consciousness and the Source, thereby allowing richer experiences to unfold, and providing some insight into the nature of the Source itself and of the extensive Consciousness"? Possible strategies are suggested, including

> openness to alternative interpretations of experience; invocation of interdisciplinary metaphors by which to express and reify those alternatives; surrender to resonance with those realities and thereby to their Source; recognition and acceptance of uncertainty as an intrinsic characteristic of both the Source and the Consciousness, and thus as an essential ingredient in the creation of any reality; and relinquishment of "either/or" mental duality in favor of creative complementarity of concepts, especially those of intention and resonance, and of Consciousness and the Source themselves.[8]

Such quasi-religious flights of fancy might not be necessary in order to understand precognition (although the strategies might be useful), at least the kind of foreseeing where we gain advance knowledge of situations we later experience in person or even reported from others. If it's difficult to see how one brain could possibly decode the specialized one-off cognitive and memory processes of another (at any rate, in the absence of a Source), it is much easier to imagine your own brain tapping directly in to its future or past state.

Because our brains are constantly in flux, forming new connections and breaking down old ones, that model of precognition and retrocognition implies that contacting yourself in the near future (rather than many years off) is less difficult. This is an appealing notion, but the evidence from remote viewing and other precognitive experiments seems to disprove it. Pat Price, one of the most startlingly effective SRI psychics, predicted in some detail events that occurred after his death. Joe McMoneagle confidently lists events he foresees over the next thousand years. Perhaps these are simply imaginary scenes summoned up in the hypnagogic state that a remote viewer enters, but it might very well be that accurate forecasts can be achieved via psi even of a future that will not be known while the viewer is alive. Of course, this can be represented as evidence that we continue beyond mortal death. It is perhaps easier to suppose that the same capacity permitting us to obtain (blurry and far from reliable) knowledge from the mind of another also lets us obtain information at a great distance and from the past or the future.

But how can that possibly be the case? Are there any theories— real, solid, scientifically acceptable theories—that are able to tie together these preposterous abilities and the regular world revealed to us by conventional science? Probably the most fashionable, the most frequently quoted, as in Brian Josephson's comment cited at the start of this chapter, and that favored by Douglas Stokes, is quantum theory. Quantum psi! First, though, before we consider explaining paranormal phenomena by turning to quantum theory, I must offer a caution or two.

Back when making an urn out of mud was hi-tech, sages taught that the universe had been turned on a potter's wheel, life shaped from damp soil by the deity's hands. In an even earthier image, ancient peoples supposed that heaven and earth formed from the sprayed semen of some lusty god; later, splattered milk from a sacred cow did a more decorous job. With the rise of literacy and the oil lamp, creation became text written upon the void: "Let there be light!" Machines put in their appearance, and before you could say "Newtonian physics" everyone figured the world resembled a big steam engine. Sir James Jeans, an early relativistic cosmologist, declared that the universe was more like a Great Thought. Today there's a computer on every desktop and in every cell phone, so it's not surprising that the Great Thought starts to look like pure information. This is information with a vengeance: *qubits,* quantum information from parallel worlds.

A bit describes a single choice with two mutually exclusive answers: yes or no, boy or girl, alive or dead. It's the basis of science. Does theory A match the experimental results better than theory B? It's also at the root of everyday decisions. Shall I take that path, or this? But wait, perhaps four or five choices are available. True, but once you've chosen one path, all the others collapse into "the roads not taken." Quantum physics makes that quite literal, in an odd way that goes beyond our usual experience. When a particle of light darts from a lamp to this page and then back to your eye, it always takes the path of least action, the shortest possible route. But in doing so, according to quantum theory, it actually took all possible pathways, which scrunched together to create that single shortest trip. What's difficult to grasp about this bizarre perspective is that *everything* in the cosmos functions by those quantum rules. Underneath the everyday stolid, sensible world, true reality is this hissing, seething fury of alternatives, sometimes obliterating, sometimes reinforcing one another. To describe it, physics needs not just bits, the yes/no, one/zero binary choices of arithmetic and computer science, but *qubits,* units of information that contain *both* yes and no, one *and* zero. Wonderfully enough, as Arthur Koestler liked to point out,

Schrödinger's key quantum equation uses the Greek symbol *psi* for the function representing all these superposed alternatives.

You can gain the superficial impression from popularizations that quantum theory's *psi* is as easy as *pi*. Consider the charmingly misleading jacket copy for David McMahon's 2006 book, *Quantum Mechanics Demystified: A Self-Teaching Guide*. The packaging is entirely hilarious. Begin with the cover and its child-simple CAPITALS:

Fun FORMAT makes this complex subject EASY to GRASP

The back jacket promises:

LEARN QUANTUM MECHANICS AT WARP SPEED
Now anyone can master the basics of quantum mechanics—
without formal training, unlimited time, or a genius IQ.

On the second page of the preface, however, things get a little more realistic: "There is no getting around the mathematical background necessary to learn quantum mechanics. The reader should know calculus, how to solve ordinary and partial differential equations, and have some exposure to matrices/linear algebra and at least a basic working knowledge of complex numbers and vectors. Some knowledge of basic probability is also helpful." By page 3 we are already into the blizzard of integrals, superscripts, and subscripts that fills the next few hundred pages. The fun, entertaining aspect of the book is presumably found in sentences like this at the end of the third page: "You can see from this formula that as v gets large, its [sic] going to blow sky-high. Worse—if you integrate over all frequencies to get the total energy per unit volume, you will get infinity."

Quantum theory is *not* easy. It is not intuitive, and it is not fun, unless you are a super geek (not that there's anything wrong with that, of course). Luckily, good simplified introductions to the ideas and historical development of quantum mechanics are freely available in dozens of popularizations, especially accessible in Wikipedia.[9] It *is* mind-boggling. I vividly remember the delirious excitement of reading one of the first books for the nonspecialist on this unhinging idea, *The Strange Story of the Quantum* by Banesh Hoffman (1906–1986), who had worked with Einstein. Hoffman was lucid and occasionally groaningly whimsical. One chapter was titled "The Atom of Niels Bohr," describing the early work of the great Danish quantum theorist, followed by its overthrow, "The Atom of Bohr Kneels." Nowadays these notions are familiar, if still unnerving, from a hundred *Discovery* Channel programs: matter is not really solid, but composed of unimaginably tiny clouds of energy, simultaneously waves and particles and neither, in some strange way resisting definition, so an electron might be found snugly orbiting near the central proton of a nearby hydrogen atom or inside a star on the far side of the galaxy. There's no way to know both the exact energy of a particle and its precise position, and not just because you have to push it around with your probe in order to study it; the uncertainty is built in to the nature and structure of reality, and that's the law. Particles can be so unsure of themselves that they tunnel right through a barrier and pop out the other side.

As the decades passed, it only got worse. In the last quarter of the twentieth century, scientists finally demonstrated that entangled particles speeding apart at the speed of light remained spookily connected. Although relativity theory insisted that no information could be passed between them faster than light—and relativity was a doctrine just as formidably supported by experiment and logic as quantum theory—the moment you learned the state of one particle you could be perfectly sure of the state of the second, even though neither state had been established in advance, even though they were both in fact and in principle absolutely random. This is not your

Grandfather Newton's physics. In a universe like that, who could say what was truly impossible? Even if the universe is "only" a gigantic mechanism, it's an indeterminate, nonlocal, and sometimes entangled quantum mechanism, not the kind that frightens spiritualists who fear they'll get their heads jammed in the gears.

So perhaps, people started to wonder, quantum theory held the key to psi. It seemed all too inevitable, in a Jungian-synchronicity sort of way. As noted, the Greek letter *psi* is the key symbol in Schrödinger's equation. There it represents what's known as the *wave function*. That is the indeterminate condition of any system before it collapses into a single final state—when, as is often said, the system is *observed*. According to quantum theory, given a state's initial conditions, it persists in a condition of *superposition*, a sort of overlap of all possible outcomes. A particle of light (photon) leaves the fluorescent light overhead; strikes your book by following a crisp, straight trajectory; and bounces from the page to your eye in another shortest possible path. In reality, though, according to quantum theory, until that photon activates a cell in your retina, its path can be regarded as a fog of probabilities, none of them actual, all of them potential and interacting, the photon also heading sideways out the door and off into space, simultaneously burrowing into your ear, making its way toward the carpet, and looping around the lamp, every conceivable possible pathway taken, most of them obliterated (the mathematics of waves assure us) by being out of phase with one another.

It seems an absurd conceit, treating this mathematical convenience as if it were real, as if it were the true, deep, realer-than-real reality. And yet quantum theory (at its appropriate level) is the best way we know for understanding and predicting and even controlling reality. It is quite fantastically accurate, in those cases where its equations are simple enough to be calculated all the way through. (Famously, it correctly predicts the magnetic moment of electrons to the eleventh decimal point![10]) The reality quantum mechanics represents is what governs the lighting of the stars and the pixels on your computer monitor. Ultimately, it is the secret of what holds atoms together and what blows them apart, the mystery underlining how your stomach digests a hamburger and how your breath moves in your lungs. It ordains the helical recipe of your DNA and accounts for those stray high-energy particles that slam through it, setting up cancer mutations in luckless cells or, enormously rarely, accidentally devising another advantageous step forward for the species.

ை

Alas, quantum theory has been turned by some enthusiasts of the paranormal into a sort of surrogate for magical chanting or invocation. One version that is increasingly encountered, depressingly, is the pseudo-quantum song warbled by smiling new age irrationalists. Careless scuba divers are prey to a disorienting buzz: nitrogen narcosis. This addlement was known once by a beautiful, alarming name: "raptures of the deep." For most of us, untrained in the sciences, advanced physics is a realm no less alien than the deep ocean. Venture beyond the safe shores of the high school lab and your common sense is put at risk. Like victims of nitrogen narcosis, you may find yourself slipping into irrational elation or despair. Call it, especially in its new age or astro-babble forms, *raptures of the shallow.*

Science is the motor of our age and a large part of its driving ideology. While it is distressing that so many of us are frightened, or dismissive, or simply ignorant of this ornate and fecund landscape, even worse is its gleeful colonization by slipshod thinkers for whom equations are mantras. For decades, a key icon of shivery mystery was the black hole, that mysterioso realm found deep in interstellar space where the laws of physics get their comeuppance. Hijacked kicking and screaming from the tensor-encrusted pages of impenetrable physics journals, black holes became sacred sites for "holistic" ninnies blissfully gobsmacked by big ideas but too lazy to work at the hard details, or too eager for effortless epiphanies. Exploring psi puts us, unless we are very cautious, in danger of this unlovely fate, this black holism. Quantum theory lends itself to such appropriation, being grounded in Schrödinger's link between observer and observed, a linkage that classic physics overlooked. Quantum observership can be hijacked, though, into a plea for the special preeminence of consciousness in "creating reality." A quantum observation is most rationally seen as any definitive interaction of a system with another ensemble of atoms that produces *decoherence.* The mutual impact of all the whirling particles that comprise dice bouncing on a table *observe one another,* so to speak, by banging into one another, jumping hither and yon, leaking away from the surface of the box into the external world, absorbing heat and light and sound from the laboratory.

Dr. Joop Houtkooper from the Center for Psychobiology and Behavioral Medicine, University of Giessen, Germany, who coined the generic name "observational theories" for the psi models we shall discuss below, described quantum decoherence thus: "Through interactions with the environment the information required to produce the characteristic quantum interference effects is almost immediately lost in the many degrees of freedom of the environment. Decoherence theories assert that this produces a definite state in the measuring instrument immediately after interaction with a quantum system, although the formalism does not tell us which state has thus become definite" (2002). Decoherence theory states that when a single particle mysteriously passes through two slits and interferes with its own possible pathways to create wave patterns on a film, it is the *film* that performs the observation, *not* the lab technician who develops it.

Or must the quantum observer stand altogether outside the system? And what observer would that be, finally? God? The Source? Or—and here's the rub—perhaps just a different *kind* of thing from dice or an atom? A *mind,* let's say. An *observing mind.* A *consciousness* that, in some unexplained way, is dualistically free of the constraints of matter and energy, able to "observe" in a locally godlike fashion. This extreme possibility, followed rigorously, led physicist Dr. Eugene Wigner to conclude that human consciousness is the crucial determinant of reality. Wigner was not drummed out of the halls of science. He was given a Nobel Prize in 1963 for contributions to nuclear physics (although not for this particular idea). For Wigner, the universe did not really exist before the emergence of intelligent life. But this metaphysical leap might not be needed.

Some prefer the quantum many-worlds hypothesis. This hair-raising doctrine—a favorite of Stephen Hawking (most famous for his acerbic best seller *A Brief History of Time* and the heroic tragedy of his medical condition) and quantum computation genius David Deutsch—holds that the famous principle of quantum randomness does not reflect an inability to decide atomic facts, but registers a "splitting" or reduplication of the universe (or at least part of it) each time a micro-level choice is determined. This is a far cry from the crypto-animism of, say, the rogue quantum theorist Fred Alan Wolf. Here is one Wolfian gem I especially relish: "The photons in [a] laser tube are bosons and so tend to 'psychically condense' into the same state. This 'boson condensation' is the physical manifestation of a universal and very human quality—the feeling of love." So laser light is driven by . . . *love! Profound,* dude. "The human body-mind is that autonomically functioning aspect of spirit, or 'qwiffness' ["qwif" is Wolf's coy acronym for "quantum wave function"], which is ultimately the form and body of God."

Black holism of this sort is mushy theft of rigorous, limited, testable ideas from science. Granted, science is based on metaphor, like all discourse, but that fact should not license the warping of its hard-won findings by those who wish to shore up their angst with a miasma of quanto-babble. Discussing John Bell's nonlocality theorem, the late

quantum theorist Heinz Pagels noted scornfully: "Some recent popu-
larizers . . . have gone on to claim that . . . the mystical notion that all
parts of the universe are instantaneously interconnected is vindicated.
. . . That is rubbish." You'd never know it from the happy zeal with
which black holists wave the Bell theorem like a mantra. And of the
ensemble theory of quantum reality, Pagels asks: "And didn't John
Wheeler, one of the physicists who helped develop this many-worlds
view, finally reject it because, in his words, 'It required too much metaphysical baggage to carry around'?" Introducing a magisterial biography of quantum pioneer Niels Bohr, Abraham Pais wrote: "I hope
that the present account will serve to counteract the many cheap
attempts at popularizing this subject, such as efforts by woolly masters
at linking quantum physics to mysticism." *Wu Li Masters,* indeed!

A recent disheartening example of this temptation is the best
seller by Lynne McTaggart, *The Field* (2002). McTaggart's Web site
claims excitedly: "Her new book *The Field* tells the story of a group of
frontier scientists who discovered that the Zero Point Field—an
ocean of subatomic vibrations in the space between things—connects
everything in the universe, much like the Force in *Star Wars.*" The
Source, the Force. What's going on here? McTaggart's writing is good
journalism, but her ambition is excessive and intellectually promiscuous, and the general perspective, alas, is laughable where it's not
irritating. She paints the customary straw-man portrait of mainstream science as "Newtonian," a term of abuse among black
holists—as if biologists and cosmologists (and the designers of the
computer on which she wrote her book) are unfamiliar with
quantum theory—which seems to me disgraceful. For example,
McTaggart repeatedly dismisses modern biology as Newtonian,
although it has been importantly quantum mechanical since double
Nobel Laureate Linus Pauling applied quantum mechanics to chemistry between the 1930s and 1950s. Consider her confusion in
extolling Hal Puthoff, Bernard Haisch, and the others for devising the

zero-point field theory, or ZPF, without noticing that it blends both *quantum* and *classical models* in a way that's closer to being Newtonian than post-Newtonian. Granted, Haisch has written in his book *The God Theory* (2006): "There exists a background sea of quantum light filling the universe and that light generates a force that opposes acceleration when you push on any material object. That is why matter seems to be solid, stable stuff that we, and the world, are made of. So maybe matter resists acceleration not because it possesses some innate thing called mass as Newton proposed and we all believed, but because the zero-point field exerts a force whenever acceleration takes place." Yet what Haisch calls "Newtonian" here is pretty much what black holists seek most ardently: innate mysterious forces and vibrations.

A notorious crux for classical science was the so-called *ultraviolet catastrophe* that threatened physics until Max Planck made his brilliant leap to a quantum explanation, which defined the initiation of quantum science; describing this heroic period, McTaggart confuses "quantum" with "classical," and Niels Bohr where she means Planck. She earnestly presents the views of German biophysicist Dr. Fritz-Albert Popp and his followers, who teach that life is crucially associated with the emission of *biophotons* (radiation from living cells that few other biologists have noticed, perhaps akin to Wilhelm Reich's mysterious "Orgone energy"). We learn that Poppian global cellular intercommunication is needed to explain how regeneration occurs in some animals, rather as Rupert Sheldrake figures they need "morphogenetic fields" to guide their development. Yet for more than half a century we've known that *every single cell in the body* (aside from red blood cells) contains the full DNA recipe, however masked some of its elements are, capable of being reactivated by appropriate chemical signals. The advent of Dolly the cloned sheep should have made this clear.

With approving excitement McTaggart cites immunologist Dr. Jacques Benveniste's homeopathic claim that cells can be affected by bathing them in the "frequencies" that are allegedly associated with certain molecules, which sounds like the old-fashioned pseudo-science of radionics to me. All of this could be true, despite the French scientist's persecution by The Amazing Randi and other CSICOP stalwarts, but it seems extremely implausible. The same applies to Karl Pribram's suggestion that mind and consciousness are located (as in my novel *The Dreaming*) *outside the brain,* this time in an extended zero-point field. Pribram's claim is that memory, as well as the workings of individual consciousness, is somehow dispersed in this mystical field, the nervous system being no more than a Hux-leyan valve or gate to perception. If that were true, why evolve senses and a modular coordinating brain at all? This is nineteenth-century Theosophy by other means, about as probable as claiming (say) that we derive our nourishment from cosmic rays, and eat and breathe only by coincidence, or for aesthetic purposes. McTaggart manages to claim that the ZPF is the explanation for everything parapsycholo-gists have found, while giving no consistent, principled reason for thinking so. This rah-rah sales-pitch element seems to have entirely taken over her own LivingTheField.com Web site, with its "48 part course in holistic living."[11] On her book's closing page, she points out that Roger Nelson and the other psi explorers "were all Christopher Columbus and nobody believed what they'd returned to tell. The bulk of the scientific community ignored them, continuing to grip tightly to the notion that the earth was flat." Actually, Columbus mis-takenly supposed the earth to be about one-third its real size, some-thing the scientific orthodoxy of his day knew and derided him for. Still, while he thought he'd found India, his error allowed him to stumble upon an entire unexpected continent. It is possible that even the easy errors and absurdities of psychic black holists might lead to a similar happy serendipity.

❧

Theories of Psi   185

You recall *your* personal experiences rather than *mine*. When I feel an itch and raise my hand to scratch my nose, your arm does not lift into the air instead. We seem to be almost entirely firewalled by our skulls. If there is a holistic channel entangling us all, it is clearly very weak and noisy. Much is made in recent psi theorizing of nonlocality and quantum entanglement, of the strange quantum connections from present to future and back to past that perhaps wind everything together (a model of quantum theory that Dr. John G. Cramer, a physics professor at the University of Washington, calls the Transactional Interpretation). Skeptics are inclined, learning of such models, to think a little more charitably about a *certain* kind of holistic hyperconnectivity. The late cosmologist Sir Fred Hoyle suggested that time loops allow future actions and states to influence the past events that, to our time-trapped egos, gave rise to them. Oddly enough, this is by no means as absurd as it sounds, and might help account for the time-reversed causality of precognition. Theoretical work by Dr. Kip Thorne and his associates have shown how reversed-time trajectories are possible without toxic consequences. Psi skeptic John Archibald Wheeler himself, astonishingly, proposed that "the universe, through some mysterious coupling of future with past, require[s] the future observer to empower past genesis."[12] And all of those well-designed precognitive experiments at Princeton, Edinburgh, and elsewhere also seem to require such time-bending effects (despite Wheeler's outrage).

Can the future influence the past that gives rise to it? We seem instantly lost in a paradoxical loop. If the future is predetermined already in the present—like corn-planted fields beyond the hill we are still climbing but can't yet see—then it would be possible to know it in advance. *But not to change it.* On the other hand, if a fixed future does not yet exist—if the word "future" is quite unlike the word "past," which also does not exist but whose influence is everywhere—then we cannot "receive information" backward in time. How could we? For there would be absolutely nothing beyond the brow of the hill—nothing but formless fog, or perhaps an infinity of overlapping possibles. So much common sense tells us. Still, the most recent century or so of scientific history has been a long series of disagreements with common sense, all of which science has won. In a 1987 Behavioral and Brain Sciences symposium on psi, one of the elder statesmen of quantum theory, the late Olivier Costa de Beauregard, offered these rather cryptic words of comfort to the believers in precognition: "[It] is now very well understood that physical irreversibility is . . . 'factlike,' not 'lawlike.' In other words, entropy or probability decrease or advanced waves are, strictly speaking, not suppressed, but strongly repressed at the macroscopic level—somewhat as antiparticles are with respect to particles. Therefore, in either field, appropriate protocols can discern the rare among the trivial."

This is precisely the claim of empirical parapsychology—that its experimental protocols in precognition (and other paranormal phenomena) isolate and amplify the rare anomalous bits of data sought among the trivial noise of random occurrence in the world. Costa de Beauregard added: "So, in terms of 'telegraphing in spacetime,' acting in the future and seeing into the past are not compulsory but customary." Most philosophers, even most quantum theorists, continue to dispute this account. We might not have a firm theoretical understanding of why the world runs downhill and not up, they tell us, but by gosh it *must* be true, or we'd be swamped in causal paradoxes. The distinguished British analytic philosopher Antony Flew, contributing to the symposium, dealt summarily with the question in the wonderfully flat-footed prose typical of his kind:

> The most damaging failure [among parapsychologists] to appreciate the essential nature of causality has . . . been the admission of precognitive ESP and retrospective PK as possible sorts of psi. For, if and insofar as both are to be defined

in terms of 'backwards causation,' this admission becomes the admission of a conceptual incoherence. Any 'backwards causation' would have to involve either making something to have happened that in fact did not happen, or making something not have happened that did in fact happen. But, if this is not self-contradictory and incoherent, then nothing is. So in these cases, although anomalous and statistically significant correlations may indeed have been found, these correlations most categorically cannot point to causal connections.

It seems a knockdown argument, until you take a look at the range of weird and terrible theories about cause and choice that modern science employs routinely when dealing with the very small and the very great. We're back to Hugh Everett's relative state or many-worlds model beloved of cosmologists, in which it is supposed that each observation of a probabilistic quantum event literally splits and reduplicates the cosmos (or part of it), each new alternative universe—including the one we're in—defined by its clean-cut adoption of just one out of the very many states the event could have taken up. On the face of it, this is the most extravagant suggestion ever seriously advanced by scientists, and indeed sometimes it is advanced by extravagant physicists, as the key to Life, the Universe, and Everything. Yet it is also admired and used to increasingly impressive effect by such undoubted geniuses as Stephen Hawking and Murray Gell-Mann (who devised the quark model of atomic nuclei). Coupling the inordinately great of cosmology with the incomparably small of quantum mechanics promised to explain the very genesis and precise nature of the universe. In June 2006 Hawking and his colleague Thomas Hertog at CERN (the European Organization for Nuclear Research) proposed a "top-down" cosmology: that the universe "began in just about every way imaginable (and maybe some that aren't)," according to a report in *Nature*. "Out of this profusion of beginnings, the vast majority withered away without leaving any real imprint on the Universe we know today. Only a tiny fraction of them blended to make the current cosmos, Hawking and Hertog claim. That, they insist, is the only possible conclusion if we are to take quantum physics seriously. 'Quantum mechanics forbids a single history,' says Hertog."[13]

John Cramer has proposed that quantum measurements combine two real phenomena usually treated in mathematics as useful fictions. Popular books on quantum theory often tell us that to get the probability of a particle's attributes (location, momentum, etc.) from the quantum equations, we multiply the quantum wave function by itself. That is, we square it. The reality is a little more complex. Wave functions employ *complex numbers:* values containing *i,* the square root of minus one. Quantum probabilities in the real world are found by multiplying a complex number (the wave function) with its *complex conjugate,* another number where *i* is replaced by minus *i.* The outcome is always real (no square roots of minus numbers), and always positive: a probability in the range 0 to 1. This mathematical shenanigans, according to Dr. Cramer, has a literal meaning. The original wave is an "offer wave" that flows out into the universe in the customary fashion. Its complex conjugate is an "echo wave," a time-reversed pulse that meets the offer wave and modulates it, creating in the encounter those quantities we observe in the laboratory.

Here's a touching parable for the Internet age: Lonely Joe sits at his computer, logged on to "Horny Friends with Benefits." He posts his self-description, his sexual tastes, and the best photo he can either find or Photoshop from an image of Hugh Jackman. His bytes blast out into cyberspace. Hundreds, thousands, tens of thousands of people read his self-advertisement. Fifteen lonely women, 283 opportunistic men doomed to disappointment, and 1,006,774 spammers instantly reply. But in this parable the first answer (from Saucy Sal, as it happens) reaches him, mysteriously, *at the very moment* he presses *Send.* He accepts the offer—and all the other possible recipients magically drop out of the story. Joe and Sal hook up and live happily ever after, at least until the next time one or the other of them logs on to "Horny Friends with Benefits."

Is there a quantum mechanism here for cutting fresh paths through the many probability worlds of the Everett cosmos? Suppose an evolving wave function with its multitude of superposed states represents many possible states of the particle, each with its own variant likelihood. Back from the myriad futures boom overlapping echo waves, to intersect and "actualize" or "collapse" the wave function in the instant of observation. Might it be, if this picture makes sense, that the moment of interaction *now* can be modified by future decisions, as in delayed-choice experiments? (Those are mind-boggling events where a decision on the means of detection that is made after experimental particles are already in flight can decide the kind of outcome subsequently observed.) Might time's locomotive thus be shunted to a sidetrack in the vast ensemble of futures, altering our experienced world's a priori probability?

Quantum weirdness, therefore, looks to some people suspiciously like psychic weirdness: instantaneous nonlocal connections, reversed-time aspects to causality. The most extraordinary aspect of this discussion, for the hardnosed lay realist, is surely that it is taking place at all. Contemporary physics, confronted by these phenomena, can no longer simply assert without argument that signals to the past are impossible. Indeed, it might be forced to accept them as routine. Causality paradoxes must be argued around very carefully, sometimes with surprising results. Standard quantum theory might be correct in insisting that retroactive effects cannot transmit useful information, but next year's science might find sneaky ways to sidestep this objection. Dr. Cramer was excited in 2004 by experiments conducted by Dr. Shahriar Afshar, at New Jersey's Rowan University, that seemed to show both wave and particle effects at the same time (http://users.

rowan.edu/~afshar/). Few prohibitions are sacred in physics for long. In June 2006 Cramer participated as well at a symposium on reverse causality sponsored (perhaps to the outrage of John Archibald Wheeler) by the eighty-seventh annual meeting of the Pacific Division of the American Association for the Advancement of Science, at the University of San Diego. "To say that it's impossible for the future to influence the past," stated convenor Dr. Daniel Sheehan, a University of San Diego physics professor, "is to deny half of the predictions of the laws of physics."[14] If that's how things really are, perhaps precognition is the advance apprehension (via, say, echo waves) of a "most probable" sheaf of alternative, decoherent histories. A future-selecting choice—which, horrifyingly, can be seen symmetrically as a past-changing choice—made on the basis of that probabilistic advance knowledge would redirect "this universe" into a less probable alternative state (one of the many available).

Might something similar account for psychokinesis, if a quantum observer—the human brain—somehow generates "echo waves" of just the right sort into an unstable system, causing its wave function to collapse in a possible but unlikely way? Or shifts the probability of a diode emitting too many pluses rather than minuses, or of a table levitating as its many jittering molecules momentarily "collapse" into an overall upward vector? For no matter how improbable anything is in the real world, its description lurks there at the margins of the superposed quantum states of its components. On the other hand, that very improbability argues against the explanation. Some have suggested that ultraviolet photon emission from the human body might somehow trigger changes in the surrounds of a psychic operator, rendering an object unstable, but how could that work? If an object being signaled (not *forced*) were so configured structurally that it *could* modify its center of mass, position, or other relevant parameter, this might make sense. But that is rarely the case with glasses and bottles resting on a table, let alone the entire table. You can open a garage door by pushing a button and sending out a bleeped message, but such tools have been designed with articulated components that utilize available dedicated power sources to alter their configuration.

⊘⊘

Still, considering psi as a quantum effect shows that Flew's armchair refutation of precognition is not nearly as definitive as he hoped. Even more dramatic was the finding, now some twenty years ago and subsequently explored with some eagerness by world-class relativists, that spacetime "wormholes" (if they exist in reality) can permit a limited kind of backward time travel. Professors Kip Thorne, Sidney Coleman, Stephen Hawking, and Igor Novikov of the Space Research Institute in Moscow are among the luminaries investigating this line of thought.

Causality is at the heart of science, despite the indeterminism built into quantum theory. So how can wormhole time machines operate without snarling into causal loops? Physicists invoke a principle of *consistency*. Any physical system must be self-consistent, including future influences reacting backward in time. Don't imagine that this conclusion is reached lightly, or that its consequences are intuitively evident. Novikov and others have described the mind-bending antics of a ball falling through a wormhole and colliding with itself. Picture it this way: a billiard ball is falling through space. There is a wormhole up ahead. Abruptly, a second billiard ball looking exactly like the first one hurtles out and strikes the first ball in precisely the right way to fling it into the wormhole. Inside the deformed spacetime, its time vector is twisted backward into what is technically called a closed timelike curve, and it falls out of the hole and collides with its earlier self, which is knocked into the hole and out of sight, closing the loop. Now the "second" or, more strictly, later ball can proceed on its altered journey through space.

But what happens if the later ball *misses* the earlier ball? Well, nothing—it can't happen, because consistency would be destroyed. You might as well ask where you would be if your mother had never met your father or, perhaps, in these days of in vitro fertilization and donor sperm or eggs, if your parental units had failed to provide their genetic material. In some cases, two or more alternative outcomes may each be consistent. Which occurs? The paradox is resolved (unsurprisingly) by regarding each ball as if it were governed by the laws of quantum theory. *Each* outcome has a certain probability of occurring, which we can write down as an equation, and what actually happens will follow from just these probabilities—as long as they are mutually consistent.

Of course, chances are, such wormholes are physically realistic only at the ultrasmall dimensions of elementary particles, rather than at the size needed to send yourself back into the past with a pocketful of old coins and a yellowing newspaper file of race results, or to a size that is large enough to fly a UFO through. Still, that need not make the possibility negligible. Some mathematical models invoke the presence of wormholes as a foam linking every part of the universe, and perhaps past and future as well. For this reason, it seems to me not altogether impossible that information-rich wormhole channels might be tapped, somehow, by perceptual psi, to bear back information patterns (if this can be done while retaining consistency) from the future. It is even conceivable that quantum uncertainties might permit forecasts of a high-probability "virtual" future that nevertheless can be altered in the present, by introducing a nonlinear, chaotic "butterfly effect"—a notion explicitly pioneered by Ray Bradbury many years ago in science fiction.

But we are yesterday's future and tomorrow's yesterday. Does this mean we and our world are now being altered by psi experiments that are already over and done with, or not yet started? In answer to such certainties, one can only shrug helplessly and point again to the many-worlds interpretation of quantum theory, which portrays people in every conceivable alternative universe, mostly very strange indeed, gazing about with smug conviction, certain that how it is with them is *how it is*, period. Is there any experimental evidence to support these rather alarming speculations? Well, naturally. Even aside from the delayed-choice quantum experiments, this is where we came in—with the parapsychological evidence for precognition. Still, we might reasonably inquire into that evidence itself. How strong is it? Does it stand up to the kind of skeptical scrutiny directed at the ganzfeld psi data? Indeed it does.

So impressive were the results of a first meta-analysis into ganzfeld that Charles Honorton and his colleagues drew together all the forced-choice experimental precognition experiments reported in English between 1935 and 1987, publishing their findings in the December 1989 *Journal of Parapsychology*. The combined results were impressive: 309 studies contributed by 62 senior authors and their associates, nearly two million individual trials made by more than 50,000 subjects. (In a properly conservative culling, all the experimental work of both Rhine's chosen but subsequently disgraced successor, Walter J. Levy, and S. G. Soal, once a famous specialist in time-displacement psi tests, was excluded; both are known to have cheated in at least some experiments.) Overall, the cumulation is highly significant—30 percent of studies provided by 40 investigators were independently significant at the 5 percent level. Yet this was not due to a suspicious handful of successful researchers: 23 of the 62 (37 percent) found overall significant scoring.

By the same token, admittedly, this means 63 percent *failed* to show significant psi. But statistician Dr. Jessica Utts made an important point when something similar had been found in the ganzfeld meta-analysis. This is absolutely crucial to our understanding and evaluation of parapsychological evidence. (Pay attention: there will be a pop quiz.) If one hundred studies are done, averaging as many as thirty-eight correct calls instead of the twenty-five due to chance, then, surprisingly, we should only expect to find among that one hundred "about 33 [statistically] significant studies . . . and a 30% chance that there would be 30 or fewer!" Here's why: The scattergun variance that arises simply from chance would mask most of the extra correct calls. This fact would remain in force *even if the respondents were picking up their extra hits through hidden radio receivers* rather than psi! It's just what happens with the statistics of phenomena that have low power.

Many critics, Utts noted, might not consider a psi paradigm successful if only thirty out of one hundred studies gave significant results. Yet they would be entirely mistaken if that were their only complaint. So, too, with the Honorton meta-analysis: a 37 percent success rate is not the failure of replication it appears to be at first glance—it is in fact very much what we *must* expect, given such a modest phenomenon as psi, and the comparatively small number of trials in most precognition experiments. Well, could this 37 percent success rate be due to the "file drawer"? Hardly. Honorton's estimate required forty-six unreported chance-level experiments for each of those in the meta-study, including those that themselves gave no significant support for the paranormal hypothesis. It seems highly unlikely that such a trove of dull experiments exists (though, undoubtedly, *some* studies of this kind must have evaded publication, especially in the early days before it was policy to publish unsuccessful experiments as well as successful ones). Nor were the results due to an excessive contribution from a few specialist parapsychologists doing so many precognition studies that their non-scoring rivals were swamped. Strikingly, if all the investigators "contributing more than three studies are eliminated, leaving 33 investigators, the combined z [number of standard deviations found] is still 6.00"—with an associated probability of chance coincidence of somewhat more than one in a billion.

The individual effect sizes were all over the place, so Honorton and his coauthor, Diane C. Ferrari, unceremoniously threw out all the studies with unusually large deviations from the mean. Perhaps this lost them some psi superstars, but it also cleaned out possible sloppy work. It might also be expected by skeptics to get rid of the supposed evidence for psi, but no: "Outcomes remain highly significant. Twenty-five percent of the studies (62/248) show overall significant hitting at the 5% level." Maybe the quality of studies explains the persistence of apparent anomalies? This, at least, was Hyman's claim in the ganzfeld meta-analysis (though he abandoned it when taxed in detail by his critics). Again, no; if anything, the significance of results *climbed* as quality improved. In other words, the better the studies, the higher the scores turned out to be. What's more, the "effect size" had persisted over more than fifty years. This measure compensates for the different sample sizes in various studies; technically, it divides the z score by the square root of the number of trials in each study. Critics have claimed that early parapsychology triumphs were due wholly to sloppy technique, and that the purported effect must fall away as loopholes were closed. Not so.

This is quite a striking result. So it seems reasonable to proceed on the assumption that precognition is not provably absurd on either philosophical, quantum mechanical, or empirical grounds. Might the eerie quantum observer be necessary after all, in both psi and ordinary consciousness?

6

QUANTUM LEAPS AND PRATFALLS

To explicate the physics of the interface between mind/consciousness and the physical brain, we shall . . . show why it is necessary in principle to advance to the quantum level to achieve an adequate theory of the neurophysiology of volitionally directed activity. The reason, essentially, is that classic physics is an approximation to the more accurate quantum theory, and that this classic approximation *eliminates the causal efficacy of our conscious efforts* that these experiments empirically manifest.

—Jeffrey M. Schwartz, Henry P. Stapp, and Mario Beauregard

On August 17, 2006, seventy-year-old Dr. Evan Harris Walker died at Harford Memorial Hospital in Maryland. He was buried next to his parents in Birmingham, Alabama, in September. A dignified obituary praised his polymathic talents: "He was a physicist, an artist and an author. He was the founder and the director of the Walker Cancer Institute in Aberdeen. He received his Ph.D. in physics from the University of Maryland in 1964. He wrote more than 100 papers that were published in scientific journals and popular magazines. During his career, he made major scientific contributions in the fields of astronomy, astrophysics, physics, neurophysiology, psychology and medicine. He held a dozen patents." What was omitted, tellingly, were his contributions to two enterprises likely to attract opprobrium: nuclear weapon design and testing, for the U.S. Navy's Aberdeen Proving Grounds, and parapsychology.

In his book *The Physics of Consciousness* Walker noted that "characteristics of QM [quantum mechanics] imply that an observer's thoughts can affect an objective apparatus directly, which in turn implies the reality not only of consciousness but of psi phenomena. I have written several papers saying that such a feature of QM is not a fault, but rather represents a solution to problems that go beyond the usual purview of physics. Thus, I have developed a theory of consciousness and psi phenomena that arises directly from these bizarre findings in QM." A more elaborate book-length treatment of his quantum theory of psi was completed when Walker died, but most of his ideas are already well known to the anomalies community—and disdained by some, perhaps because most parapsychologists are not well schooled in physics, and possibly because the remainder are. (The book's back jacket blurb is not by some leading neuroscientist or quantum theorist; it's from Dr. Harvey Cox, of Harvard's Divinity School.) It is fair to say that Walker's attempt to blend psi and quantum physics was a landmark effort, but it's far less clear how persuasive it was.

His thinking about the relationship between mind and external world was focused by a quantum paradox published by Eugene Wigner in 1967. That, in turn, had spun off analyses of state collapse by John von Neumann and quantum pioneer Erwin Schrödinger. Human beings seemed to emerge from the science of quantum theory as observers of a world we do literally *create* or at least harvest (in some bizarre but fundamental fashion) in the act of observing it. The heraldic beast for this revolutionary philosophy was a ghost split between realities, a creature neither dead nor alive, known in the trade as Schrödinger's Cat. This luckless feline was the imaginary victim of a deeply troubling thought experiment posed by Schrödinger. The hypothetical animal is confined in a sealed box—*perfectly* sealed from the rest of the world, which is a key to unlocking the eventual paradox—with a cyanide capsule that may or may not be smashed open by a poised hammer triggered by the decay of an unstable atom. Because the quantum physics of radioactive emission is governed by probability, rather than the kinds of direct cause-and-effect logic we are used to, it can be arranged that there is, within the period covered by the experiment, on average just one chance in two that the isotope will decay and emit an electron. If it does, a detector registers the quantum event and the hammer falls. Instantly, the lethal gas suffuses the box and the cat perishes at once. If there is no decay, the cat is spared. What could be a simpler choice of outcomes, or a more sinister one? When we open the door, we'll find either a snoozing puss, or a pitiable bundle of cold fur and bones.

Obviously. But what was happening to the cat *while the door remained sealed?* Many people find their minds seizing up at this point. Open or shut, the cat is either poisoned or not, either dead or alive. Not so, argued Schrödinger, who thought he was performing a sarcastic *reductio ad absurdum* but instead kindled decades of perplexity. The poor cat's fate is intimately linked with the state of the radioactive isotope inside the sealed box, which—quantum theory assures us—remains *both* intact *and* decayed, in a condition of probabilistic suspension, so long as no measurement is made of the state of the box from outside its confines. Against all orthodox reason and common sense, that radioactive atom inside the box subsists as a mixture of overlapping or superposed waves. One wave stipulates an intact atom, and that tiny, microscopic state of affairs becomes magnified up to the macroscopic realm, where we live, *as the palpably alert cat.* The other wave specifies a decayed atom, and in this case, because the atom and the rest of the prepared box are muddled together, its microscopic state is amplified, sadly, into a dead cat. The wave functions of both possible outcomes are *entangled*. Neither is sufficient alone to describe the condition of blended box, isotope, hammer, flask of cyanide, or cat until the instant when the system is "observed."

At that moment (but which moment exactly? observed by whom, exactly?), the superposed waves *collapse* into a single, definitive description. The whole awful experiment stops being probabilistic and snaps into certainty, with a probability of one point zero. The infinitesimal effects at the level of a single atom, of its protons and neutrons and orbiting electrons, are correlated with the health of the cat, and since both states of the cat are entangled and overlap, the animal has been *neither* dead *nor* alive. Until observed. What's needed, apparently, is an interaction from *outside* the box. But then where can this mad cascade stop? Eugene Wigner concluded that human consciousness is the crucial determinant of reality, but asked: what if a friend of the cat observer observed the scientist in turn? The first scientist, after all, is just another swarm of quantum particles, a set of superposed descriptions awaiting state collapse *from the outside*. So Wigner's friend collapses both scientist and the cat, but who, then, causes the collapse of the state vector describing Wigner's friend?

You can chase this string all the way up to the entire universe, regarded as "a cat in a cosmic box." Stephen Hawking and his colleague James Hartle have written down their best equation for the wave function of the universe, but who collapses it? Nobody, they tell us. Wrong question. Meanwhile, Sir Roger Penrose, another of the world's most brilliant mathematicians, argues for a different answer, one where quantum effects that are harnessed by the structure of microtubules inside neural cells invoke the force of gravity to bring about state vector collapse. Sir John Eccles, 1963 Nobel Laureate in brain neurology (the same year Wigner shared a Nobel for physics), joined one of the great philosophers of science, Sir Karl Popper, in claiming that soul couples to matter in the brain via just such quantum effects. Like those celebrated experts, but with his own jerry-rigged model of brain and modified Schrödinger equation, Walker wanted *consciousness* to observe reality's overlapping alternative pictures, and he wanted *will*—active human choice or volition—to collapse them. And he thought of a clever psi experiment to test it.

If psi is not limited in space (as many parapsychology experiments seemed to confirm), then by the physical principle known as Lorentz invariance it should not be limited in time, either. The slightest acquaintance with anomalies research confirms this as well, since precognition delivers results as good as clairvoyance—in other words, remote viewing—in real time. But that restricts psi to its cognitive aspect, *knowing* in advance. What if psychokinesis—psychic *doing*—also traverses space and time, allowing you not just to modify the stream of ones and zeros now coming out of a random event generator sitting on the table, but also reaching backward into the past and skewing its output last Thursday?

This seems a horrendous possibility. It brings to mind the desolate child's cry: "Make it didn't happen!" Obviously, we think, this is self-evidently absurd. Even if a cause can occur after its effect (although, as we've just seen, physicists insist that you need a black hole or wormhole to make it work), surely you can't eat your cake and have it, too. Literally. If the cake was sitting on the table last Thursday, and you gobbled it down greedily, surely there is no way you can reach back into the past and contact yourself or your friend sitting at the table to insist that you put the cake down and seal it away safely. After all, last week you observed the cake, you observed yourself eating cake, and so did your friend. Walker did not disagree with this. But suppose instead, blindfolded, you had reached into one of Suitbert Ertel's opaque bags and drawn out, one by one, ten of the fifty numbered balls and placed them in a second opaque bag for safekeeping. Better still, suppose you had a mechanical device driven by a quantum randomizer to draw out the balls and hide them away.

A week later, you carefully record your intention and place your-self in psychic rapport with those hidden numbered objects—not now, where they still remain securely locked away, but back at the time that they were selected randomly. Bring your will to bear, and insist that each of those drawn balls is number four. Nobody knows whether or not the draws were entirely random, nobody has observed the balls. Now, having attempted to modify the selection in the past, you fetch the balls and observe them. But wait. This is not the same as a Schrödinger observation, according to Walker. If he's right, last Thursday those balls were already observed, by you; now, *psychically.* You count them. Four out of the ten are numbered "four." Remarkable. It seems that you have retroactively modified a random outcome.

Here's another possible implication, a variation on the decline effects mapped by Dick Bierman. If the order of time is open in this fashion to manipulation by psi, it does not follow that the exhaus-tion of a particular psychic protocol must necessarily lead to an observed decline from past to present to future. If psi manifestations in any given protocol become eroded or exhausted with repeated use of that protocol, perhaps an interesting test could be contrived by running numerous unobserved sets of trials, then retroactively psi-observing a random sample of one-third or one-fourth of them, obtaining scores and effect size, then repeating until all data have been observed and checked. If erosion is real, we might expect the time series to show no *overall* decline, because the decline will be scattered through the sequence. (It would take an awful lot of work, though, and a lot of delayed gratification.) Presumably such psi ablation would increasingly contaminate efforts to gather and examine pre-existing ordinary experimental data—for example, clinical fMRI studied belatedly for signs of presentiment. This might create an inter-esting curve, in which the farther back one went in search of obscure sources of preexisting data, the less evidence there would be in them of psi, creating an overall *incline* effect, at least up to the present day.

This is the kind of mad thing that seems to follow from what are now called observational theories. Dr. Joop Houtkooper defines them thus: "Observational theory can be best characterized by its central axiom: **'The act of observation by a motivated observer of an event with a quantum mechanically uncertain outcome influences that outcome.'** . . . A 'motivated observer' [has] a conscious or unconscious preference for one specific outcome over the other possible outcome(s)." Bierman and others have subjected such theories to experimental tests during the last several decades, usually using random number generators with their tedious, all but endless streams of ones and zeros. The finding appears to support the effectiveness of psychokinesis even when it is applied retrospectively, through the barrier of time.

But something even creepier turned up during these experiments. You can set a *pseudo*random number generator to pour out a babbling sequence that is indistinguishable from chance, but is entirely determined by a complex algorithm plus a seed number, which starts the sequence going. Every time you put the same seed number into the RNG—say, on an ordinary desktop computer—you will get the same output. So what happens if you try to bias such an output retrospectively? When this has been tried, the results show the same kinds of deviation from chance, over the preordained length of the experimental run, as when you use a true quantum random number generator. But this is insane! Has the psychic force, or whatever it is, modified the algorithm inside the computer? The best answer seems to be that *selection of the seed number itself* is what was modified. And that, in turn, implies that the goal-seeking nature of psi has the extraordinary capacity to predict the outcomes of numerous possible seed numbers when amplified by a given algorithm, match them against the specified bias, and then choose the best-fitting seed number to be made real. On the face of it, this implies contact with some sort of immense cosmic calculator, perhaps Jahn's Source. Alternatively, quantum physics might come into play once again. Perhaps when the

seed number was itself selected by random process, your world comprised a series of superposed states with utterly predictable, deterministic outcomes, which psi is somehow able to filter or assess, finally
selecting the single state most appropriate to your desire, leading to
state reduction or collapse of the state vector. If something like this is
the case, it will call for a radical reconstruction of how we understand
the world. But, then, so might everything concerning psi.

Walker was happy to accept the results of retroactive psychokinesis as evidence in favor of his observational theory. But as a scientist, trained in the use of measurement and extrapolation, he could
not leave it at that. After all, his primary goal was to solve the measurement problem—the baffling question of exactly when, and how,
the deterministic but multiform outcome of Schrödinger's equation
abruptly collapses into one single state, which we experience as the
real world. Walker knew that the brain and nervous system can't be
decoupled from the mind. Whatever is going on with psi, and with
measurement in general, must in consequence be a function of the
way the brain is built (or vice versa).

His research grew elaborate, exploring the microstructure of the
brain, the way neurons and synapses couple together, neurotrans-
mitter chemicals pouring back and forth, activating receptors, being
absorbed, the whole inconceivably complex orchestration of the cere-
bral symphony. His results are remarkably concrete and specific.
Consciousness, he concluded, operates at the rate of roughly 10^8 bits
per second, while will (the psi channel) only manages a lowly 10^4 bits
per second, and speech less than 100 bits per second. (Elsewhere, with
admittedly unwarranted precision, this becomes 47.5 million bits per
second for the consciousness stream, 58.7 thousand bits per second
for the lower will stream, a ratio of roughly 1,000:1 (2000). Because
of this disparity, psi is only able to modify or bias the world by one
part in somewhere between 1,000 and 10,000 in a random stream of
events—the latter figure, interestingly, comparable to the average

overall results found during the two and a half decades of the PEAR work with random number systems.

What's more, in order to describe this process in traditional quantum mechanical terms, Walker was obliged to introduce a new term into the Schrödinger equation. This kind of tinkering is not unknown to the history of science. When Einstein understood that the speed of light in vacuum was the absolute limiting velocity permitted by relativity theory, he modified the traditional simple Newtonian equation of motion by adding the so-called *gamma* term, which in words rather than symbols reads: "1 divided by the square root of 1 minus velocity squared over the speed of light squared." Walker's suggested modification to Schrödinger's equation has met, however, with rather less general acclaim than Einstein's suggestion.

This cruelly denuded sketch of Walker's elaborate theorizing deserves the rebuke he offered a critic: "A rigorous presentation is necessary if one is to obtain a technically satisfactory understanding of the theory and not merely an abstract" (1984). Let's consider a slightly technical summary, without equations, where Walker notes that while the classic way to represent information is a written inscription, we can also represent information

in terms of the potential states in quantum mechanics that can come into being as the outcome of material interactions— the potential state of an atom that becomes excited by radiation as they strike it, for example. The condition of such an atom is given by the transition probabilities that can be calculated from the state vector that represents the excited atom. The state vector description of the atom as given by quantum mechanics, therefore, contains information about the physical system it describes. . . .

A third kind of information is also at work in the brain. . . . One potential state becomes the actuality, the observed

event. One synapse does fire. Something does happen. It is
the selection of which event will occur, out of all that
could occur, that is the third type of information at work in
the brain. . . . Thus we see that mind actually has two parts:
consciousness and will. (2000)

How did his theorizing stack up against alternative models?
Walker claimed that it provided no less than thirty-nine tests of both
consciousness and psi, none of which had failed. Mere coincidence?
"In parapsychology, we discover 'observer effects' as a general way of
talking about psi phenomena. In physics, we encounter the 'observer
problem' in which QM imposes the incredible conclusion that the
observer has an effect on QM systems. This simply cannot be coincidence! This must lie at the heart of the solution to the problem of psi
phenomena; and, indeed, an understanding of psi phenomena and
consciousness must provide the basis for an improved understanding
of QM" (1984).

One of the main sources of experimental evidence for Walker's
analysis came from physicist Helmut Schmidt's indefatigable work
with random number generators. That had also been developed on
the basis of thinking about quantum mechanics. Dr. Schmidt initially
offered what he called a "teleological model," akin to Rex Stanford's
goal-controlled PMIR, in which all observers of a particular random
phenomenon contribute to its final realization. It begins "with an
ensemble of all possible world histories with their corresponding
probabilities as they would be predicted by conventional physics"—
that is, by the Schrödinger equation. "And, to introduce an element of
noncausality, the model postulates a psi law that modifies the probabilities for the different world histories. This law is teleological in the
sense that . . . the outcome of a coin flip depends on its implications
for the future history of the world. This is certainly implausible, but so
is precognition . . . the model does not claim to 'explain' psi. It does

not even try to discuss what happens inside a subject's head" (Schmidt, 1984). Importantly, lest conventional physicists be more scandalized than necessary, Schmidt notes that this model "changes only the probabilities of world histories that were already possible in the frame of conventional physics. . . . Only statistical laws are affected."

One troubling implication of this model, which Schmidt admits candidly, is the *divergence problem:* "The teleological model, with its spacetime-independent structure, provides for all [observed] noncausal effects and arrives at PK, precognition, and the other forms of psi from one basic mechanism. This attractive high degree of symmetry leads to a divergence problem in the sense that future observers obtain an unreasonably high influence on the present." After all, if one person has the power to *observe* the past with its multitude of possible histories, and thereby have an impact on the choice of a single history, why shouldn't other observers as well, scattered in time and space? It seems an intractable problem, once the idea of an absolute causal time order is abandoned. Walker's observer theory, by contrast, insists that only one initial conscious observation can collapse the state vector—even though it can happen in the future after the collapse! Evidence to date is weak but does suggest that perhaps Schmidt's more promiscuous approach is closer to the truth.

Dissatisfied with this model, in 1982 Schmidt proposed a quantum collapse explanation not altogether unlike Walker's. Again, we return to Schrödinger's overlapping "ghost" states of any given system prior to observation (whatever *observation* means). When a system undergoes quantum collapse, settling into a single definitive state, that situation can be described by a number of different interpretations. Schmidt's version introduces a gradual reduction or exponential decay of the ghost states, tuned by what he calls an "alertness parameter." Putting aside the technical details once again (this is not a textbook, after all), it's worth noting that if this model is correct, psychokinetic scores produced by two or more separate subjects working together or consecutively can't be greater than the best score achieved by one of them alone. Thus, the divergence problem is solved at a stroke. The particular virtue of the earlier teleological account, for Schmidt, was that it requires no concept of consciousness, whereas the quantum collapse model demands an effect by a conscious observer. Which model is to be preferred? "For the physicist," Schmidt writes, "the ultimate goal of psi research would be the discovery of some novel microscopic law of Nature of great mathematical simplicity and beauty, from which all psi effects could, in principle, be derived. That law would qualify, from the physicist viewpoint, as an 'explanation' of psi" (1984). Let's not be too hasty, though, he cautions. First we need to accumulate sufficient nuanced evidence, building a basis "for a later, more complete understanding."

In 1977 J. W. Hartwell had already suggested that in any psi experiment "all humankind may be participants to a degree that depends on the *probability* of the individual's later becoming an actual observer of the outcome of the experiment" (Bierman and Debra Weiner, 1980). That is, not only are the past and present dependent on the entire future in which observations might be made, but they also depend on *potential* future observers! Early tests of this vertiginous model involved the destruction of a large amount of data immediately after their observation, as feedback, only by the subject or participant. The degree of success in these results was compared to the scores gained when the experimenter subsequently observed the intact results. Dick Bierman and Debra Weiner reported that their first results contradicted Hartwell's expectation. A quarter century and more later, Bierman told me:

> I certainly think that observational theories in some form have the best potential. But we shouldn't be blind to the empirical facts that there are instances where feedback seems to be less than relevant, thereby undermining one of the

basic premises of most observational theories. Hartwell's "potential observational theory," where the role of actual observation is replaced by the probability of a result to become observed, easily accommodates instances of apparent irrelevance of feedback. Joop Houtkooper's work on multiple observation (either repeated and/or with multiple observers) is important to shed light on the theoretical issues. However, I am also almost sure that the observational theories are not the final answer.[1]

The teleological model is itself a kind of quantum theory, depending on the superposed alternatives all hanging at right angles to one another inside the Schrödinger equation. For Dr. David Deutsch, quantum theorist at Oxford University's Clarendon Laboratory, every event is literally doubled and reduplicated, with small crucial variations, in trillions upon trillions of diverging parallel realities, spread through infinite lateral time. (Although he is open to the possibility of a form of time travel through these alternative realities, Deutsch is a fierce foe of psi claims: "It is utter rubbish.") If you're having a bad hair day, there is another version of you, in a universe at right angles to this one, who's doing just fine. And there are a zillion more, taking up every conceivable alternative position in between. Some of them are dead. Some are on Mars. A few are sharing Graceland with Elvis, who is married to Princess Di or perhaps John Lennon. This remarkable postulate, as we have seen, was not freshly minted by Deutsch; the many-worlds hypothesis was proposed in 1957 by Hugh Everett III, who in this universe died decades ago. Until fairly recently, though, nobody had pursued these striking notions with such ruthless intellectual attack, following them all the way down.

For Seth Lloyd, a distinguished theorist of computation and professor of mechanical engineering at MIT, and a pioneer of quantum computing, it's crucial that everything in the universe is made of qubits. Every atom, every elementary particle, is a tiny quantum computer that registers information, processes it, and passes it along not as a stream of crisp binary code but in a blur of superimposed but tightly constrained possibilities. Quantum computers can scratch itches that ordinary digital computers can't reach. As yet, in labs only very limited quantum computers have been built—but they do exist, so we know that they're not just somebody's clever but unlikely brainstorm. With a quantum computer you can explore simultaneously all possible answers to a given question and see the correct answer instantly fall out as the incorrect answers obliterate one another—like out-of-phase sound waves in a first-class passenger's noise-reduction headset. Where are these computations being performed? In all the superposed probability states of the computer simultaneously.

But if the entire universe is not just a regular computer but instead a cosmic quantum computer, what is it calculating? Dr. Lloyd has offered an inevitable answer: it is calculating itself. The universe is no longer a book written by a divine author, scribbling and discarding multiple drafts. It is a colossal computation in which all those drafts come into existence at once and, in the jargon of physics, interfere constructively and destructively with one another. In the end, what you see is what you get, but it's just the faintest after-image of the compressed multiplicity of its computation. "In this picture," Lloyd says, "the universe embarks on all possible computations at once."

Deutsch, too, insists that his outrageous version of reality ought not surprise us, or at least physicists, since it is simply the best theory available to science (quantum mechanics) taken perfectly literally without metaphysical evasions. He has argued brilliantly that something as ordinary as the splash of illumination shone by a flashlight on a wall cannot be explained without acknowledging that a sort of "shadow light" from many other adjacent universes is leaking into our own, interfering with the beam we see by blurring its bright, sharp disk into concentric rings. In those universes, naturally, a few photons from our local universe also leak across, fritzing *their* beams of light. "It follows," Deutsch notes, "that reality is a much bigger thing than it seems, and most of it is invisible." That multitude of overlapping histories is needed to explain some of the most basic features of our ordinary world. Deutsch argues ingeniously that gene sequences performing the same function in different adjacent worlds must be closely similar, while "junk DNA" littering genomes will vary at random. It is precisely this consistency in functional informational structures that marks them off from noisy rubbish, however complex and elaborate the noise might seem. Could psi (pace Deutsch) be accessing such shadow alternative worlds, or steering the observing psyche through a sheaf of them in search of heart's desire, or its closest handy approximation?

Later, I'll suggest diffidently that one of psi's critical evolved functions might itself be directed toward the maintenance and even the improvement of the genome. If Deutsch is correct, perhaps the presence of locally selected "fit" genetic modules in many adjacent worlds might reinforce one another, strengthening the probability that they will be realized. Strictly speaking, this might seem impossible, since the formalism of superposed states specifies that they are orthogonal to one another—that is, at right angles in as many dimensions as it takes, mutually unreachable. But then so, too, should be the shadow photons Deutsch invokes to explain quantum interference. In 2004 Dr. Paul Davies noted that quantum mechanics "provides an explanation for the shapes of molecules, crucial to the templating functions of nucleic acids and the specificity of proteins. The Pauli exclusion principle ensures that atoms and molecules possess definite sizes, which in turn determines not just templating, but differential

diffusion rates, membrane properties and many other important biological functions. Quantum mechanics also accounts for the
strengths of molecular bonds that hold the machinery of life together
and permit metabolism."[2]

More speculatively, Davies asked whether "the qualitatively distinctive properties of life owe their origin to some aspect of 'quantum
weirdness.'" For example, if living things retain some of the aspects of
the microcosmic quantum realm, processing information with qubits
rather than bits, perhaps having access to the fantastic exfoliation of
quantum computing that poses its questions to a multitude of alternative superposed states, this might explain how life searches for good
strategies with surprising speed and success. But as Sir Roger Penrose
and his colleagues have realized, such appropriation of quantum
powers at the microcosmic level run up against the problem of decoherence. Cells are hot and noisy, environments that swiftly destroy
quantum coherence. *Swiftly* hardly does the pace of the thing justice:
the entanglement of the particles is destroyed inside 10^{-13} seconds.
Davies suggested two ways that living cells might protect their onboard
quantum computers from collapsing in this way. One is to shield them
with molecules that act as heat pumps; the other depends on a phenomenon that is weird even for quantum mechanics, the quantum
Zeno effect, which rather resembles the folklore claim that a pot does
not boil if it's watched. These mysteries are best explored in Davies's
own paper. What struck me was his comment:

> Though it is easy to believe that quantum superpositions
> might accelerate the "discovery" of a specific, special, phys-
> ical state (e.g., the "living" state), there is an element of tele-
> ology creeping into this mode of thought. *We* might be
> familiar with what it takes for a system to be living, but a
> molecular mixture isn't. The concept of a "target sequence"
> or "goal" at the end of a search is meaningless for molecules.

Nevertheless, a quantum search could speed up the "discovery" of life if there is some way in which the system "knows when it is getting hot," i.e. if there is some sort of feedback that senses the proximity to life, and focuses or canalizes the search toward it.

This is precisely the problem novelist Greg Egan faced in trying to rationalize an evolution accelerator in his inventive novel *Teranesia*. There is no optimal outcome in evolution, only a nonstop tussle between existing competitors and the genetic material (including rare mutations) available for organisms to build into their offspring. The giraffe doesn't know that a longer neck would favor its child, even if it knew how to bring that about by Lamarckian onboard genetic engineering. The same objection applies to Davies's living quantum computers—*unless* there is a way to gain advance access to (possible) future outcomes. And that might be the key function of psi: to search the many pathways opening from this moment into the future, using precognition; to test any solution found against those discovered in competing genomes, both here and in alternative realities, by clairvoyance or remote sensing; and finally, to tweak the genome of which it is one expression, by anomalous perturbation (in this circumstance of literally mindless cells, "*psycho*kinesis" is a particularly ill-formed term).

Indeed, it is possible that the origin of life itself forms part of a "psychic" temporal loop, with conditions in the future reaching backward into the multiple possibilities available in the past and selecting, by observation, those conducive to its own existence. This might be no more preposterous than the standard explanation for the origin of the universe: that it quantum-tunneled from a state of nothingness into being, *just because it could*. Davies concludes: "The deep problem of biogenesis—the discovery of a pathway from non-life to life—may be advanced by viewing the origin of life not so much as a chemical process, but as the emergence of a specific, coded, information-processing system from an incoherent molecular milieu. . . . Advances in quantum information theory promise to cast this problem in an entirely new light, and could point the way to the decisive breakthrough in explaining one of science's deepest mysteries." With psi thrown in as an extra empirical factor currently ignored by almost all scientists, decisive breakthroughs might very well be in the offing.

But do these quantum-based theories—or models, if that is putting the matter too strongly—offer the best and most convincing available explanation for psi? And perhaps equally important: what is it for?

7
IF IT'S REAL, THEN WHAT?

We do not have positive definitions of the effects that generally fall under the heading of anomalous mental phenomena. In the crassest of terms, anomalous mental phenomena are what happens when nothing else should, at least as nature is currently understood.

—Edwin C. May, Jessica M. Utts, and S. James P. Spottiswoode, "Decision Augmentation Theory: Toward a Model of Anomalous Mental Phenomena"

Let's stipulate for the remainder of this book that the reality of psi has been sufficiently established, and that some of its piquant properties and characteristics have been mapped, at least provisionally. If you remain adamantly skeptical, come along for the ride anyway. Think of it as an entertaining exploration in the mode of science fiction: *if psi actually existed, what would it be like?* What would be its purpose, range, and limitations? Could it provide fresh and disturbing insights into the nature of the universe, of life, of mind, perhaps of some spiritual realm recognized so far only by faith and not by science? Or must it be nothing more than an oddity at the margins of physics, always escaping the attempts of technologists to harness it for use, rather as the mysterious but ubiquitous characteristic of quantum nonlocality resolutely refuses to lend itself to instantaneous communication across great distances?

One problem is the very terminology we must use about these phenomena, a problem built so deeply into this discussion that it is scarcely visible. A term like "paranormal" is a confession of ignorance, as is "extrasensory perception," although not quite to the same extent. These are ways of saying only that something is happening but we don't know what it is. And that's alarming, because these days most of the time when something is happening experts can provide us with a pretty damned good idea of what it is and why it happened that way. Granted, in some misleading sense that has always been the case. If your neighbor gave you a dirty look and your buffalo, goat, cow, or chicken then got sick, experts stood ready to assure you that this plight was a result of sorcery, almost certainly wrought by that very neighbor. If you didn't know any better, it was a convincing explanation. These days we do indeed know a great deal better, even though it's not polite to say so. But there is a peculiar kind of humility or abstinence in using a term like "paranormal." We might not know what is causing such phenomena—alas, we almost certainly don't know—but we're not completely in the dark. Or rather, even if we are in the dark, even if we have no idea at all where the light switch is or what function it serves, we do know that "psi" is a *positive* term, a label for something that happens to us or for something that we do to the world.

Something actively brings about the congruence between Zener card and psychic call, or between unknown target and remote viewing perception. It might be that whatever supports the process is mediated by some energy field (or, more generally, *some condition*) outside the control and awareness of the psychically functioning operator. Consider, for analogy, the rather strange experience known as "blindsight." It has been known for more than thirty years that some completely or partially blind people—that is, those with at least one part of their visual field that simply does not support vision—nevertheless have the eerie capacity to make accurate guesses *identifying what they can't see.* You might hold up a card showing an *X* or an *O* and, without their having any subjective experience of *seeing* what is displayed, the blindsighted will "guess" correctly far more frequently than they should be able to do by chance. Are they using psi? No. Are they lying about their blindness? No. The explanation lies deep inside the brain. Neuroscientists now know that vision is a complex construct, not a simple, transparent glass pane on the world. It starts with preprocessing at the retina and passes from one staging post to the next until each image is recognized and coordinated by the visual cortex. The first important staging post (known technically as *V1*, or the striate nucleus) can suffer damage, destroying the experience of vision. Even so, lesser and alternative pathways can continue to function, passing information from the eye to the controlling cortex *but without any accompanying experience of sight!* Perhaps this is a little like the way we all detect certain subtle odors at a subliminal level, not consciously recognizing them but still reacting emotionally, sometimes bristling when we meet a new person, sometimes feeling strangely attracted, without ever knowing that we're picking up a few stray molecules of pheromone scent from that person's glands.

Imagine, then, an entire community of severely blindsighted people, H. G. Wells's Country of the Blind, perhaps. Suppose this condition has become genetically inherited rather than being the result of rare individual brain damage. The cliché is that the one-eyed man is king in that country, a cliché that Wells delighted in upsetting. Certainly any throwbacks who still possessed normal vision would seem to possess an incomprehensible advantage, and might choose to retain it by refusing to explain the situation to the blindsighted. For the majority with blindsight (rather than true blindness), life would have certain curious aspects. At some times during their endlessly dark experience, they might keep bumping into furniture left carelessly in their path, while at others they would seem to have a kind of mystical ability to avoid such bruises. After a time, they might begin to make the connection between warmth on the skin and this uncanny sense, although by no means would that always be the case. Sometimes when it's dreary, cold, and raining, they'd find that their guesses still took them in the right direction, although perhaps less convincingly than during the hottest part of the day. Sometimes there might seem to be no connection whatsoever, especially when they're indoors, aside from some mysterious clicking sounds. Because they don't understand about illumination by sun and moon, they can't even begin to guess that a few genuinely sighted people switch electric lights on and off.

So psi might be a bit like that. Apparently, the blindsighted have no reliable sense of when their guesses are accurate and when they're simply random. So, too, intermittent "illumination" by some pervasive but intermittent psi medium might provide no opportunity for training. Given the small effect size found in parapsychology experiments, most correct guesses must be a result of chance coincidence. Consequently, immediate reinforcement of the inner states associated with generating those guesses will tend, ironically, to create what's called an *extinction paradigm.* That is, the more you attend to the way you feel as you make correct guesses, the more often you'll be noticing completely irrelevant and distracting states of mind. Attempts to train people to recognize when psi is active have traditionally been ineffective, perhaps for that reason. Adding "confidence calls" in association with guessing at targets has not, by and large, offered much evidence that people know when they're doing it by psi rather than by chance—or rather, in most cases, by following the patterned templates that the human brain uses as a rough-and-ready approximation to randomness. (An exception was reported in the *Journal of Parapsychology* by Bierman in 1999, where ganzfeld subjects rating their confidence at

40/40 scored at 37.8 percent rather than the 25 percent expected by chance. Those who were confident at less than 39/40 averaged only 33.3 percent right.[1]) It does seem that excellent gifted remote viewers like Joseph McMoneagle have indeed learned how best to place themselves in the appropriate psi-conducive state, and have developed interpretive protocols for extracting the maximum amount of paranormal information from the all-but-random jitter that their brains produce in that state. And if the LST peak and lowest effectivity times are sustained by further investigation, we might be able to home in on whatever is giving rise to these auspicious opportunities (or to circumvent whatever is blocking psi most of the time).

But perhaps the key to psi lies not only in our environment, but also in ourselves, as blindsight is due not just to illumination by "invisible" light but also to the deep visual-system structures of the brain that remain intact. One of the key storytelling devices in old science fiction tales was the idea that psi is a *wild talent* (a term borrowed from Charles Fort, an inveterate hunter after the unusual and the absurd), a skill that might in the future be roused to activity in mutants born in the aftermath of atomic war. Wild talents might be a new evolutionary development, or the recovery of one suppressed in the ancient past. Many science fiction novels and short stories explored this topic, especially in the heyday of the great science fiction editor John W. Campbell, whose *Astounding Science Fiction* magazine, later known as *Analog*, made a fetish of the paranormal during the 1950s. Only a few hundred thousand ardent fans ever read this digest magazine, but I suspect that most of our cultural images of "paranormal powers" derive fairly directly not from folklore or superstition but from Campbell's feverish editorial suggestions to his ingenious writers. That idea of mutation either creating new and mysterious abilities or facilitating the reemergence of old capacities muted or lost to the pampered and civilized crowded into their cities is based in a very crude misunderstanding of how genetics actually functions. But you can't blame the writers or readers for that; most of these literally astounding stories were dreamed up before Francis Crick and James Watson had uncovered the secrets of DNA and its coding.

Now we do know a great deal more about both the genome and the ways it expresses itself under the selecting forces of the environment, only to be subject itself to selection. Theoretical turf wars continue to rage over the fine-grain detail of evolution, but among those who have studied the matter professionally there is no doubt that the capacities passed down from one organism to its offspring are shaped, pruned, conserved, and selected by several key factors. The main one, as Richard Dawkins has pointed out repeatedly, is that each of us is the offspring of a line of gene-prescribed survivors. Most of the creatures born during the last three and a half thousand million years have perished without leaving any direct lineage. Granted, many die by what amounts to accident, however able they might be in other circumstances (making "survival of the fittest" something of a misnomer; "survival of the fit and lucky" might be more appropriate). Famously, all the untold billions of dinosaurs that covered the planet and many of the creatures under the sea were obliterated in one ghastly tragedy at the end of the Cretaceous era, 65 million years ago. Was it the random work of a small asteroid or huge meteorite slamming into the ocean, poisoning the planet against large-scale life for hundreds of years? Perhaps, despite popular opinion, it was instead a terrifying tectonic event, a series of immense volcanoes blowing sunlight-blocking debris into the atmosphere. In any event, the particular comparative fitness of one dinosaur species compared with another—or, more importantly, of one dinosaur compared to others of its own species—suddenly counted for absolutely nothing. And that dreadful global killing field left room for the small scurrying creatures that developed, over the ensuing tens of millions of years, into the mammals that today dominate the world, notably us.

But most evolutionary change is on a far smaller and more dispersed basis. Here, indeed, point mutations to single genes, or simple sexual recombination of stable genes, modify the recipe, define each new individual, and are then passed down in turn as small modifications to that individual's offspring. Only very rarely does a mutation or recombination result in a really drastically different progeny, although this ceaseless mixing and remixing of genes in the given lineage does account in part for the variation we see in the children of a family. There's a general overall family resemblance, but some of the kids grow to become taller or shorter, heavier or lighter, somewhat smarter or somewhat duller. Except in the case of developmental defect in the womb, or childhood illness, your offspring will tend to resemble you considerably more than they resemble anyone else in the street or in a nation halfway around the world (or if you're a male, so you hope). If there are genes that have the effect of installing and controlling psychic abilities, we would expect this long, slow natural process to shape the pattern of inheritance through generations. Some critics brandish this insight with glee, asserting that if psi appeared by mutational accident (rather than chance recombination, swiftly swept away), it would burn through a population within only a few dozen generations, driving everyone in the direction of superb precognition, telepathy, psychokinesis—yet obviously, they argue, this has not been the case. Ergo, psi is a snare and a delusion.

Many questions are begged by this assertion, as we shall see. For a start, psi might not be an *adaptation,* like wings that evolved from limbs, but what some evolutionists call an *exaptation:* the conscription of spare capacity for some totally novel purpose. The notion was proposed by Stephan Jay Gould and Elizabeth Vrba, and not all evolutionists accept it. Gould offered an amusing analogy, comparing such captures with "spandrels," the wedge-shaped corner sections of old church ceilings that artists filled with interesting forms. The spandrels were not a deliberate part of the design of the ceiling; they were an unintended side effect of geometry, of the material limitations of the era's architecture. Arguably, the ability to do mathematics, and especially to excel in some lofty branch of the science, is a spandrel in the brain. Obviously we inherited through evolutionary selection our sharp, stereoscopic, color-sensitive vision, our capacity to walk on two feet, and our ability to grasp and manipulate deftly with our opposable thumbs. Those abilities were absolutely critical to survival and thriving in the hundreds of thousands of years during which modern Homo sapiens emerged as a sideline from our evolutionary ancestry. But ancient men and women had no need of the capacity to work through partial differential equations, let alone eleven-dimensional topology. That's probably why most of us still have quite a lot of difficulty with mathematics but can talk effortlessly about sports or politics or fashion.

In the same way, it's conceivable that psi is a spandrel—something like the way certain people are said to be able to pick up radio signals through the fillings in their teeth. No dentist ever put metal inside a filling to allow a customer to listen to the Top 40 or the latest shock jock. For all I know, this phenomenon might be nothing better than an urban legend, but you can see how a lump of metal inside the oral cavity might resonate to invisible radio waves that are constantly passing through all of us, pumped out by a million radio transmitters. I do not mean that psi is literally a kind of mental radio. That analogy turned out to be unworkable many decades ago. We know that electromagnetic impulses, with their speed-of-light limitations and their inverse square attenuation, simply do not have the muscle or the reach for the job. Some physicists, notably in Russia, made a brave attempt to find versions of electromagnetism that would do the trick—extremely long waves near the frequency of the resting brain, around five cycles per second, the alpha rhythm, capable of being passed from one side of the earth to the other with minimal attenuation inside the wave guide of the upper atmosphere—but none of those candidates has the capacity to carry information into the past, which is required for precognition.

But let's suppose for the sake of argument that psi is an evolved function, like sight, like laughter, like the capacity to digest oranges or milk. Many adult humans, as it turns out, have great difficulty digesting cow's milk. A special enzyme, lactase, is required to do the job, and most Japanese, for example, have lost the coding for that enzyme from their genome. It's not impossible, therefore, that the genetic coding for whatever it is in the human brain and body that supports the experience of psi has been suppressed or lost over the millennia or even millions of years of recent evolution. Equally, it's possible that we inherit code for building, maintaining, and using psi-supportive apparatus but that today's dominant culture fails to elicit the flowering of that capacity. We know that *stereopsis*—the capacity to see depth by combining two slightly separated images, one from each eye—is something that needs to be learned. If one eye is masked during infancy, the portion of the brain dedicated to translating its impulses into the experience of vision is actually suppressed and eventually conscripted for different purposes. It's possible that we all start out with sequences of our genetic coding bent on developing extrasensory perception and psychokinesis, but without suitable training those functions never develop—or, worse still, are actively suppressed.

Apparent evidence in favor of this view is the supposed prevalence of psychic communication and sorcery among traditional peoples who are untainted by rude reductionist materialism, television soap operas, "Western" medicine, and literacy. This has been claimed of certain Australian aboriginal cultures as well as the shamanic traditions of the remoter portions of Russia and lodges of certain Native American tribes. It's not impossible, but seems to me about as likely as claims for the superior physics and astronomy of those cultures. Humans are superstitious animals; it's a side effect—perhaps a spandrel—of the way our brains are set up to lock on to meaningful conjunctions. People believe the craziest things, and not just loony ideas. We cling to them in the face of overwhelming evidence. A streetlight shines through a certain tree and casts an image of Jesus Christ, complete with crown of thorns, on a nearby wall. A miracle! People flock to marvel and pray in the street. Cognitive scientists quietly point out that human brains are replete with specialized feature detectors. These useful nerve cells scan the world for other humans, clamoring for attention when they find anything that looks a bit like a familiar face. Blotches of shadow, in other words, trick people's brains into seeing something that literally is not there. But, like Fox Mulder in *The X-Files,* the devout want to believe. Still, the premise of this chapter is that psi actually is real. Consequently, it's not absurd to suppose that some cultures are particularly effective in provoking its appearance, even in learning to submit to its oddly vagrant and mischievous nature. The problem with trying to learn about psi from such cultures is that the very process of investigation and translation—which is almost always done within the larger context of colonial or post-colonial conquest and ruin—seems inevitably to contaminate or even destroy such traditions. It might be too late to learn much of value from these sources.

We come back to evolution and the possible functions and constraints of psi. Is it a method of seeing and hearing farther than our senses can reach? Is it a means of distracting a potential foe or hunted animal, or even of imposing force upon one or the other, the equivalent of striking with fist or thrown stone? Is it a specialized form of communication between members of a clan or family, or even between parent and child, especially mother and child? At a time when evolutionary psychology has been growing in popularity as an explanatory narrative, at least in those parts of the world where sacred creationism and intelligent design are not the favorite themes, it might seem surprising that few theorists have tried to fit the paranormal into an evolutionary framework. The sociobiology of psi! We need to step free of absurd popular iconographies and grasp the complexities of evolution. It's important to acknowledge that most people who have adopted evolution as their working hypothesis of the cosmos, as I do, flinch from the horrors its pressures routinely produce. Its strategies, as noted, are founded on the single fact that each of us is the end result of a long line of ancestors, beings who all successfully passed on their genes. Happily, not every product of random mutation and natural selection is vile. Such a filter can employ many strategies and tactics, including love, kindness, generosity, honor, and artistic and scientific genius. If psi is an evolved function, presumably it will share our better angels as well as the worse.

Evolution conserves nothing but "inclusive fitness": comparative success in making and protecting progeny by any means possible, including both altruism toward other family members or gene sharers, *and* their heartless exploitation. If a strategy benefits inclusive fitness more effectively by building minds and bodies exposed to violence, decay, and death, plagued by demons of unhappiness and strife, it will triumph over its sweeter-natured rivals. Our origin in evolutionary contest is why we suffer agony and inflict it (as well as luxuriating in devotion and joy), and goes a long way to explaining the persistence of evil. When such insights were first systematized some third of a century ago as sociobiology, there was an outcry from decent, fair-minded souls. Social Darwinism! Racism! Reductionism! Nazism! Those charges were usually wide of the mark. Indeed, most evolutionary forays into human affairs were made by compassionate academics. Today, with the Human Genome Mapping Project complete, there's an urgent need to reconcile neo-Darwinism with our tenderest cultural longings. Dr. E. O. Wilson, doyen of sociobiology and coiner of that now passé term, speaks these days of *consilience* and *biophilia,* terms with a kindlier tang to them. How is kindness possible if we are ruled by our relentless inheritance? Because genetic regulation of behavior is complex, and indirect. Although the final arbiter of natural selection is an individual's fertile progeny—or close kin sharing the same genes—that brute fact cannot serve as a sufficient moral guide. Values, created by complex human brains, have burst free of the imperatives of the genes. Alas, our hard-wired animal inheritance has not. So we live in endless conflict between what is and what ought to be (whatever we decide *that* is, this week). This fact is hardly news. The flesh is notoriously weaker than the spirit, and even the spirit has its own ways of going lethally haywire. Take a look at Iraq or much of Africa. Psychoanalysis and the arts are eager to tell us about mixed or hidden motives, most of them sexual in one way or another. Evolutionary psychology, though, is starting to show us how to track down why and how these apparent aberrations are structured into life.

In 1987 parapsychologist and academic Dr. Richard S. Broughton, author of *Parapsychology: The Controversial Science,* took the plunge in his presidential address to the Parapsychology Association. His speech was titled "If You Want to Know How It Works, First Find Out What It's For." Broughton was taking almost the first serious steps into an evolutionary account of the psi phenomena. Parapsychology has not yet managed to attract many thinkers from the hard-core evolutionary biology camp—more's the pity.

Broughton noted at the outset that parapsychologists had lately been unsettled by evidence that their experimental results might not necessarily be produced wholly by their subjects/participants. The *experimenter* might prove to be even more critical in the attainment of good or even bad scores. This "experimenter effect" had been suspected for a long time, but by the mid-1980s it was coming under close scrutiny at a time when theories developed from quantum physics raised the possibility that results were literally created by those who observe them. Noting this shift in emphasis, Broughton acknowledged a resurgence of the opinion that psi would only be understood "if we view it as a form of transcendent, interpersonal field or as a holistic system involving a large number of 'participants.'" Giving in to this temptation is, admittedly, one small step toward mysticism, one giant leap away from traditional scientific assumptions, and Broughton wished to resist its lure. Regarding psi as an ability (like hearing) or physiological function (like digestion) is the only solid basis for testing it, experimenting on it, asking that question "what is it *for?*"

Suppose the answer is, instead, that it's leakage from our spiritual nature, or a by-product of the way our immaterial soul connects to our material flesh, or a sort of unintentional side effect of metaphysical illumination (as certain Eastern traditions insist, disparaging miracles as low distractions). Then one might be enticed to sink into meditation and worship rather than to return to the laboratory and its waiting apparatus and participants, most of them students looking for credit in Psych 101 courses. But if psi is a tool of survival under Darwinian selection pressure, the methods of evolutionary analysis might offer some surprising insights.

One hazard of this approach much assailed by foes of what they regard as "reductionist science" is that explanations of this kind can have a fatally Just So flavor (along the lines of Rudyard Kipling's stories such as "How the Elephant Got His Trunk"). Sometimes this is due to the inadequacy of the explanation, its easy fanciful quality, but sometimes what is revealed is the inadequacy of its critics, who fail to grasp the elaborate mathematics of contemporary evolutionary explanations. Fortunately, we can avoid those bristling mathematics here, mostly because nobody seems to have applied any just yet to the paranormal. So if what follows can be readily maligned as Just So storytelling, let's admit that up front and see how far it can take us anyway.

Psi, as Broughton notes, is a pretty defective way of communicating and manipulating the world. If you want to send a message, use Western Union, or your cell phone. On the other hand, some commentators have stressed how psi, as a kind of ordering principle, appears to serve unconscious needs. Given the way paranormal events seem to spring out of nowhere—and, even when we try to elicit them in the laboratory, tend to show up spasmodically and unreliably—they certainly don't *seem* very orderly. But some recent experiments by Suitbert Ertel cast a fascinating light on the possibility, as we'll see shortly.

Psychokinetic psi might modify odds slightly in our favor, warping probability space, so to speak (whatever that would mean), a position reminiscent of Rex Stanford's psi-mediated instrumental response, or PMIR, model. Stanford went on to modify his own earlier model, speaking instead of *conformance behavior,* whereby the world is modified to meet individual needs. (If psi is shaped by evolutionary pressures, it is conceivable that this conformance operates even more sharply upon the genes themselves, their integrity and organization, their maintenance and even selection. Again, we will return to this speculation.) In the modern world this sort of largely unconscious ability could readily backfire, giving rise to mechanisms of suppression.

It is reasonable to suppose that the eagerness of young men to prove themselves in battle against neighbors pressing upon their resources, or simply to steal what the neighbors possess, results from an enormous selection pressure applied through a million years of hominid evolution. This readiness to snap from ordinary youthful irritability and boastfulness into murderous pack behavior is probably set by social perceptions rather than by any "instinct for violence," but when circumstances are grim, it is easy to tap and tremendously difficult to switch off. Given an evolved psychic ability to read and manipulate the world, especially including other people, it is possible that we also experience countervailing pressure to mute or disown and disable psi.

Even if that's the case, as Broughton notes, it's probable that "the primary function of psi is to help the individual survive when faced with serious threats to health and safety *and also to gain a competitive advantage in the struggle for survival*" (his italics). Of course, "survival" today means social success, survival as a successful contributor to an increasingly complex social order. Perhaps, Broughton notes, these psi effects manifest themselves in ordinary life as *intuition* and *luck.* Note that the needs psi is conceived as serving are not your conscious wishes, any more than your horny interest in the object of your attentions or your conscientious concern for your children's welfare is driven by a conscious desire to maximize the spread of your genes. What we experience as motives is filtered through a hierarchy of behavioral templates and urges, or so this model of evolutionary psychology tells us.

But since the underlying game plan is success, one likely target of psychic effort, according to Broughton, will be "a survival strategy to counteract the advantages consciousness gives to one's competitors. . . . It could well be elusive and obscure by design." This is not such a crazy idea. Lying and deception are often successful strategies, but only if the cheating goes undetected. Unfortunately for the would-be cheater, chicanery exacts its toll upon the body, and especially upon the face. We have evolved subtle detectors of deception. For example, we can usually tell by observing the muscles around the eyes when somebody's smile is false rather than spontaneous. So cheating works successfully only if the cheater starts by fooling himself or herself. Hypocrisy is a sharply honed tool in the armory of social contest. If psi were visible, palpable, and reliable, perhaps it would not be as useful; indeed, "it could get one into trouble," the sort experienced by a cardsharp caught at his wiles. Well, it's a hypothesis.

This would not be an either-or proposition. As Broughton notes, game theory specialists have shown that evolutionary strategies often blend several possible moves. Rather than being a dedicated dove or hawk, the more successful strategy is often to sit quietly, abiding by the rules as a dove 80 percent of the time, backing off when threatened, but randomly, 20 percent of the time, going ferociously feral in a hyper-hawklike fashion. Many other mixed strategies are feasible and offer superior advantages. Hiding your psychic ability from yourself most of the time, even all the time, would permit this kind of subtle use of uncanny skills. If this is the case, Broughton suggests, there is a useful lesson here for parapsychologists on the hunt for psi and methods of controlling it: "It is high time that experimenters realize that the only reliable way to get psi out of subjects is to give them reason to use it—real reason."

He quotes Ed May: *If you want to find good subjects, find successful people.* Joe McMoneagle, for example, was tremendously successful in staying alive under threat in Southeast Asia, leading others through danger by following instincts he barely understood at the time. A famous science fiction novel now fifty years old, Alfred Bester's *The Stars My Destination*, features almost every conceivable kind of real or imaginary psychic ability. At its heart is a new ability—apportation or teleportation, the capacity allegedly documented by Charles Fort of translating yourself from one place to another just by wishing it—that Bester calls Jaunting. Bester whimsically attributes this name to its hapless discoverer, Charles Fort Jaunte, who accidentally sets fire to himself and escapes a gruesome end only by teleporting himself seventy feet to a fire extinguisher. Scientists investigate ruthlessly, sealing Jaunte inside an unbreakable tank that fills slowly and lethally with water. Moments from death by drowning, he teleports to safety, dripping and coughing. Melodramatic, to be sure, but Bester might have been on to something there. Psi, if it's a survival mechanism, will respond much better to urgent tasks, or even diverting and interesting ones, while fading into the background when presented with yet another ten thousand trials at skewing a random number generator.

One brilliant if apparently wildly zany test of the PMIR model was trialed in the 1980s by a French researcher, Dr. Rene Peoc'h, using a small robot (a tychoscope) governed by a random event generator. The clever move was to conscript a genetically inherited trick that attaches chicks to their mothers. Called "imprinting," it causes the baby birds to form an instant and powerful attachment to the first moving creature they see close at hand—in the wild, of course, their own mother hen. This imprinting bent can be subverted easily by humans, who can trick the creatures into bonding with them, or indeed with friendly dogs or even machines. Peoc'h bonded his chicks on the robot. A diagram in Peoc'h's paper shows the recorded tracking of a control run in the absence of chicks, when the little robot wanders and jinks every which way, entirely at random as expected. This is contrasted with an absolutely mind-boggling tracking chart of the robot's journeys when the motion-bonded chicks watched it from their adjacent cage. The machine's wanderings are visibly restricted to one side of its arena, most movements jammed up against the wall next to the besotted chicks. Presumably, wanting the beloved machine/mother to stay near them, increasingly anxious if it moved away, they manipulated its random number generator to keep it nearby.

In evidence of this mad suggestion, other chicks that had not been imprinted produced no such anomaly in the robot's ambling. As usual, alas, this apparently watertight evidence for psi slams into the perfectly justifiable critical objection that the work was published in 1988 but carefully and successfully replicated only a handful of times (according to Dr. Gertrude Schmeidler). One group reported to the 2005 Parapsychological Association meeting that their replication attempt had failed—but they had modified Peoc'h's methods, and arguably failed as well even to achieve imprinting in their test animals. It seems a simple enough experiment for critics to repeat. Probably it never will be, at least while psi is barricaded outside the gates of orthodox science.

❧

Evolutionary behavioral science these days is hard-edged with formidable mathematics. Bull session ideas don't cut the mustard. Even so, here are some idea starters. Let's suppose that any genome that is conducive to psi might contain at least two relevant recessive genes. The instructions represented by recessive genes are masked in the presence of their alternative, dominant form (or allele). Eye color is a well-known example: if mother has blue eyes, it means she has inherited the blue-eyed allele from both of her parents. If the father of her children has brown eyes, none of her children will have her eyes, because brown is dominant. But half of them will carry a blue-eyed gene, and their own children can express it if they, in turn, mate with someone bearing a blue-eyed gene, even if it is not expressed in the parent. This sneaky now-you-see it, now-you-don't aspect of Mendelian inheritance must have caused a lot of marital strife in the past, before the ins and outs of the algebra were understood.

So in our ancient environment of evolutionary adaptation, tens and maybe hundreds of thousands of years ago, perhaps only those offspring who received a double dose of each psi gene were effectively psychic—their capacity perhaps enhanced and focused by ritual, liturgy, wild pharmaceuticals, and so on. Few of their own direct offspring would manifest the full-blown capacity. The psi-conducive genes might generally circulate invisibly, manifesting themselves only in distant cousins. All kinds of possibilities arise, some amusing and some dreadful. Suppose there is conflict between clades (groups sharing features derived directly from a common ancestor) that are separated by various degrees of expression of such recessives. The phenotypes involved—the people, I mean, their bodies and brains—need not be consciously aware of the basis of their antipathy for one another. At best, the psi activity induced by these genes might still be marginal. Still, the numerically dominant "non-psychic" or "low-psychic" elements of any community might keep their psychic minority (shamans, mystics, even some telephone psychics, for that matter) on a strict utilitarian leash, or encourage them to adopt celibacy, or periodically round them up and burn them at the stake. This latter expedient might remove the problem for a time, but recessives would still lurk in the gene pool, ready to reemerge.

But suppose psi is subtler than that. Suppose its chief evolved function is to work recursively and directly *on the genome itself*. Imagine something like a genetic sequence that manipulated/deformed its own alleles directly. Such an effect could deform the normative allele into the psi-conducing configuration. The presence of such genes might then encourage the competing selection of other genes that defend the "non-psychic" alleles, acting to suppress or obliterate the "psychic" form. Paradoxically, they might even do that by something like psychokinesis, conscripting the talent they act to suppress.

To calculate the population consequences would require serious genetic algebra. Presumably such "prionic psionic" effects would eventually attain equilibrium, either becoming self-limiting or reaching a balance by mutual contest. If not, a species might fairly quickly tend to become saturated with one form or the other. More likely, as discussed above, some proportion of the population might use a mixed strategy that is not only permissible but might well have a decisive advantage. A version of this notion was published in a 1969 science fiction story by Anne McCaffrey, "A Womanly Talent." The Talented begin to emerge in the twenty-first century and are brought together and validated with the use of "Gooseggs" (a form of "ultrasensitive electroencephalograms," perhaps resembling an advanced version of Roger Nelson's EGGs). Young Ruth, apparently without Talent, has recently married powerful precognitive Lajos Horvath, a fire hazard sensitive who advises an insurance company. During sex, remote monitoring instruments reveal in their "coital graphs . . . a tight, intense, obviously kinetic pattern." Why is Ruth expending this prodigious kinetic energy?

The answer surprises everyone in the story, but tends to strike one these days as at least somewhat sexist: "For the exercise of a very womanly talent. . . . What is the fundamental purpose of intercourse between members of the opposite sex?" Well, reproduction. With this clue, the Talented realize that Ruth's new baby manifests inheritance patterns that are possible *but extremely unlikely*, making the baby an immensely strong telepath. What Ruth did was to rearrange, completely unconsciously, "the protein components of the chromosome pairs which serve as gene locks and took the blue-eyed genes and blonde-haired ones out of cell storage. And what ever else she wanted to create Dorotea." She does the same in order to perform a miracle of healing on someone else. "She can actually unlock the genes!" cries one amazed specialist. It is a provocative idea, when generalized. Could psi be, at its deepest level, a process for optimizing the genome of the species as swiftly as possible within an environment that precognitive psi itself suggests will remain unchanged for many generations? Perhaps there is a clue here to the somewhat surprising discontinuities in evolutionary history claimed by Stephen Jay Gould and his colleagues, dubbed "punctuated equilibrium." Species can appear convulsively (on the geological timescale, that is), and then persist almost unchanged except at the micro level for many millions of years, until a new challenge, often environmental and catastrophic, leads to a new leap. Might it be that psi acts to consolidate such major adaptations, even to feel out potential pathways in evolutionary space and encourage the structure of the genome to settle into an appropriate form? A version of this idea, as we've seen, was advanced by the brilliant science fiction writer Greg Egan, in his 1999 novel, *Teranesia*.[2] Oddly enough, it also forms the basis of a radical theory—quantum evolution—suggested by a professor of molecular genetics, Dr. Johnjoe McFadden,[3] and by mathematical physicist and Templeton Award winner Dr. Paul Davies, as noted in chapter 6.

I played with a somewhat related idea in a science fiction story I published more than a quarter of a century ago: "The Ballad of Bowsprit Bear's Stead."[4] In that fiction, psychic superstars are parasites powered by the future psychic capacities of their descendants (and perhaps by the abilities of their long-dead ancestors). Presumably this psychic vampirism must be limited by some law of diminishing returns in either temporal direction. It pays the family line to be very fertile, and for all gifted individuals to exercise their own fecundity. On the other hand, I conjectured various deleterious side effects, the sorts of accumulated disorders that are common within inbreeding populations. This dynamic might seem to fly in the face of, for example, celibate saints and mystics, but it is notable that their *siblings* have traditionally been encouraged to be hyper-fertile.

Suppose that some "witch burner" gene conduces to various kinds of furtive or masked opposition to those carrying "psi genes." Those expressing this gene in behavior might not even be aware of their own gene-biased motives, in much the way, as we've seen, that effective liars first convince themselves that they are speaking the truth, the more effectively to defeat the feature detectors of their victims. At a conspiratorial extreme, one might imagine that the traditional habitats of people who retain a high concentration of the "psychic genes"—so-called primitive peoples—might come under systematic attack. Nuclear tests, one recalls, were conducted in territories populated principally by Native Americans, Australian aborigines, Pacific Islanders, and their equivalents in the Soviet East. (Obviously, I hasten to add, there are far less exotic explanations for this colonial practice: such barely habitable or remote places are where these unfortunate people were frequently consigned.)

❧

Dr. Robin Taylor lectured for a decade in psychology at the University of the South Pacific, Suva, Fiji Islands, after postdoctoral research as the Koestler Parapsychology Unit at the University of Edinburgh. In 2003 he published a preliminary attempt at combining evolutionary theory with current knowledge about psi, explicitly responding to Broughton's call. He calls his own model Evolution's Need Serving Psi (ENSP) "to denote this heavy slant towards a biological emphasis," offering an explanation for why psi is probably based on largely unconscious mechanisms. Psi is imperfect, he argues, because it scans the environment for necessary information only to a limited degree. (Rex Stanford's own original PMIR account had psi constantly scanning, but he dropped this requirement after calculating that too much energy and effort would be required for this kind of nonstop psychic radar, and because his modified model sees psi as "goal oriented" rather than reactive.) Being explicitly Darwinian, Taylor's model expects psi to function most effectively in support of any organism's family, as well as itself, and, as we've seen previously, in unpredictable settings and circumstances. If that's so, the best way to train psi would be to make use of startling tasks of some urgency, although it is fairly unlikely that the drastic Jaunte technique for eliciting paranormal phenomena would pass today's ethics committees.

But still, if psi is under evolutionary pressure like other attributes, why hasn't it improved to the point where it's at least on an equal footing with our sharp vision, or even our rather blunted sense of smell? We have seen some arguments why it might achieve a kind of sullen dog-eat-dog equilibrium, and it's apparent that evolved functions do vary in their degree of expression throughout a species. Still, psi is so feeble (at least in our current environment) that it seems an odd candidate for selection. As Taylor asks pointedly: "Why do large numbers of people still fall foul [of] disaster" when even unconscious psi would be subject to positive selection only "if it confers a higher rate of survival to an organism?" Perhaps, Taylor suggests, the evolutionary hurdle to the improvement of psychic capacity is too high to cross by the small steps available to natural selection. Alternatively, perhaps truly effective psi would be a poisoned chalice in social species like our own. In what is often regarded as his finest novel, science fiction writer Robert Silverberg's *Dying Inside* (New York: Scribner's, 1972) presents a scarifying image of a once high-functioning telepath:

> I haven't had much love in my life. That isn't intended as a grab for your pity, just as a simple statement of fact, objective and cool. The nature of my condition diminishes my capacity to love and be loved. A man in my circumstances, wide open to everyone's innermost thoughts, really isn't going to experience a great deal of love. He is poor at giving love because he doesn't much trust his fellow human beings: he knows too many of their dirty little secrets, and that kills his feelings for them. Unable to give, he cannot get. His soul, hardened by isolation and ungivingness, becomes inaccessible, and so it is not easy for others to love him. The loop closes upon itself and he is trapped within. (52)

(Silverberg has confirmed the necessary point that the novel is not, of course, really about telepathy but about the universal affliction of aging and loss of power and potency. "The telepathic stuff is just the narrative vehicle. I read some Rhine etc., but it certainly doesn't represent my own paranormal experiences, of which there have been none. The book is often mistaken for explicit autobiography. At some science-fiction convention years ago," he added with his distinctive dryness, "a woman who had just read it said, 'I didn't know you were one of us!' I replied, 'If I were one of you, you would have known it long before this.'")

What's more, if cognitive psi works by conscripting and repurposing our ordinary senses and internal imagery, as seems likely, an exceptionally vivid psychic ability might routinely interfere with the utility of those ordinary senses. As Douglas M. Stokes comments sardonically in his important review paper of 1987, "It would not do from an evolutionary perspective to have one's mind engaging in clairvoyant voyeurism of an orgy down the street while a fire is going on in one's apartment," precisely the kind of risk that Silverberg's tragic, defeated character David Selig runs even as he finds his craft ebbing. Powerful spontaneous evidence for the existence of psi is found in the anecdotes of those who have evaded death due to a prophetic vision, and yet many people continue to die in accidents that might have been avoided.

It can be argued that we resist the vagrant urgings of psi out of fear of embarrassment (after all, much of the time these "intuitions" will be mistaken), not wishing to act against habit. Taylor finds this explanation inadequate. Actually, he argues, it's mainly a consequence of built-in psychic inattention. "To avoid an overload on cognitive processing power, psi is used only periodically to 'sample' the environment. The rate of sampling is suited to the context or environment in which an organism finds itself." If a child's environment is essentially benign and predictable, the sampling rate might be set low, and perhaps that set point becomes wired in. Training psi, in any event, might require stimuli "directly related to hunger, thirst, shelter, mate selection and avoiding danger," even if these are only simulated.[5]

A more global interpretation of psi has been developed over many years by veteran American parapsychologist and therapist Dr. James Carpenter, and published in the last several years under the whimsical title "First Sight," a play on the cliché that psi is *second sight*. It rather resembles the construct of lunar astronaut Dr. Edgar Mitchell, who regards psi not so much as an ancillary as a primordial or *first* sense:

It is now reasonably certain that, even before the historical period, humans as well as nonhuman species had an internal intuitive, or "visceral," knowing, which has been called a "sixth sense." It now seems that it should have been called the "first sense" because modern research leads us to believe that it is based upon a complex form of quantum correlation that was certainly present in nature long before species evolved to their current stage, and even before the planetary environment evolved to produce the normal five senses. Research in a frontier field called the quantum hologram leads us to that conclusion. It helps to explain virtually all the "nonlocal" intuitive, psychic, and numinous effects that humans have reported during historic times but have previously been unable to understand in terms of this world's natural processes.[6]

Carpenter's model (in *JPara*, March 2005) assumes that

each organism, by its nature, extends beyond itself into the larger pre-sensory surround. Psi is assumed to be neither knowledge nor action, but to belong to the outermost temporal edge of those normal pre-experiential mental processes by which the mind structures all its experiences and commences all its actions. Psi processes are posited to function normally as the unconscious leading edge of the development of all consciousness and all intention. This unconscious functioning is normal and continuous, and is a constituent element of all experience.

British biologist Rupert Sheldrake, too, sees psi this way, as an amoeboid probing with extended psychic pseudopodia of our surrounds in space and time, testing our opportunities to survive and thrive via a conjectural and all-embracing nonlocal "morphic field" (Sheldrake, 2004). Carpenter agrees: "For this conception to be sensible, we need to assume that each organism exists, by its nature, beyond its own physical boundaries, in some sort of commerce with the larger surround of space and time."

Psi, then, gets us ready for what's coming at us, like a social guide muttering discreetly in our ear, tipping us off to the names and status of those we are about to meet, directing us to the correct dinner utensil or appropriate garment for the occasion. It anticipates our needs, provides a sort of anticipatory Google search on current and upcoming experiences, and usually brings the results to us in the form of "inadvertencies," apparently irrelevant events and experiences that nevertheless "implicitly express the action of the orienting activity." Jung would have called them synchronicities. Psychotherapists like to weasel them out as clues to what's going on in the unconscious of a client.

In short, "ESP is the leading edge of the mind's ability to move to the next experience; PK is the leading edge of the mind's ability to move the next effect to its intention. These psi processes are continuously active but normally unconscious and implicit . . . [they] are not unusual or exotic. They function as the initiating part of the mind's perpetual preconscious working toward the end of constructing its experience and framing its choices."

It's a radical solution, as bold a conjecture as physics' sea of virtual particles seething and bubbling in the background of all spacetime. Neither the alerting messages nor the first opening moves, either within the body or pushed into the outward world, get dealt with at a conscious level, usually. They provide the very context of consciousness. In this they resemble stimuli that psychology knows as "primes"—orienting experiences that make it easier to choose one path rather than another. If you are shown photographs of many unhappy faces, your mind is primed to read the next ambiguous image as expressing misery rather than, say, pensiveness (as the classic Russian film director Sergei Eisenstein discovered). This kind of priming can occur at a subliminal level, and in Carpenter's view it's largely what goes on with psi, all the time. His model is confessedly nearer to the earlier version of Stanford's PMIR, retaining the element of active scanning, although "psi is conceived as being even more broadly active and important, functioning ubiquitously as the initiating point of all human experience and volition." *Meaning* begins to suffuse the experienced universe even before we are aware of the experiences. It is the aspect of psi that German parapsychologists such as Walter von Lucadou have embraced, influenced as they tend to be by the philosophy known as phenomenology.

Because the human brain/mind is not a passive camera, but actively works on incoming data, comparing it with expectations and then constructing our experience, meaning is intrinsic to any item of perception or action. Experimental psychologists demonstrated long ago that subliminal priming stimuli are readily assimilated into an image when they are meaningful to its gestalt, but can be actively repudiated or ignored when the brain fails to detect any salience. Perhaps, Carpenter argues, the otherwise baffling phenomena of psi-missing is rather like this. When an internal construct is comparatively sharp or strongly biased, there will be less opportunity for psi-priming to modify our assessment of the situation we are in, and consequently how unconscious preprocessing will repudiate the psychic information input, rather in the way you unconsciously place a hand in front of your mouth in the moment that you repress or censor some dangerous utterance, or swiftly turn your guilty gaze away from something you are actually avid to look at.

Of course, many if not most subjects of parapsychology tests in both ESP and PK produce only dud scores, neither hitting nor missing. If we assume that psi is a constant contributor to our experience, what's up with that? Carpenter has a not very satisfactory answer drawn from empirical work in perception: "[A] chance-level score across a period of effort is not taken as an absence of psi process but as an implication of switching intention and direction of approach such that neither an additive nor a subtractive reference to the material is evident overall." Oh. Well, okay. If that seems a bit of a cop-out, Carpenter adds the now-customary insight that to enhance visible psi you need to make the target more personally relevant or meaningful. What's more, repeated guessing at a sequence of fundamentally non-meaningful targets is likely to produce boredom, unconscious inattention, and eventually aversion—that is, first of all the decline effect, and then a switch to psi-missing. Which brings us back to psi as an evolutionary adaptation, whether or not Carpenter is correct in his claim that it is functioning all the time in the undergrowth of our personal information processing.

Several years ago, on a restricted Internet discussion list on parapsychology, a young British computing specialist posted a remarkable claim in support of a general approach to salience and variety as a key to psi activation. (He prefers to retain his privacy for the moment.) He holds an arts degree and a masters in engineering from Cambridge University, and has nearly a decade's experience in the commercial world, specializing in pattern recognition and processing, computer learning, and communications systems. His private research interests have focused on how techniques applied to communication, learning, and recognition systems might be applied to psi, drawing upon techniques to improve experimental design (for example, using a maximum entropy gradient, a method developed as well by the former SRI/SAIC theorists). He wants to identify and remove any unwarranted design complexity or assumptions, including the background assumption that a given run of trials using random targets is really independent of those that precede and follow it. This is, on reflection, a curious assumption to be held by those who acknowledge that psi appears to operate freely in both time and space. You might expect a kind of blurring of perception, which in turn might be one reason why evolutionary selection pressures have rather less traction with the paranormal than with the usual, plodding one-thing-after-another.

Here is the young man's remarkable account, used with his permission, of an inventive approach to precognition:

It began as a typically naive test of my own precognitive ability; but ended up revealing (to me at least) some of the innermost workings of this phenomenon.

I started out with a fresh deck of playing cards. I took out an ace, two, three, four and five of the same suit. I shuffled them well and laid them out in front of me, backs up. OK, let's relax, calm the mind and look at the backs of the cards until some impression comes to me. After brief passive relaxation and observation, momentary glimpses or flashes of the cards came to me in their appropriate positions. When I felt comfortable that I had a strong sense of each card's face value, I turned them in order of my predicted ascending face value: ace—hit, two—hit, three—hit, four—hit, five—hit (obviously). OK, nothing spectacular statistically (120 to 1). I thought: so let's have another go.

Again I shuffled and laid the cards out; again I relaxed and passively waited for some glimpses of the cards to come to me, which they did after a short interval. The difference this time was that the glimpses seemed to be slightly confused. Each glimpse of a card seemed to have an additional ghost image of one of the other cards superimposed. Still, each glimpse did have a dominant face value. Slightly less confident than in the previous run, I turned over each card in order of ascending face value; ace—hit, two—hit, three—hit, four—hit, five. Pretty good, I thought; good luck comes in threes? I shuffled and laid out the cards. Again I passively waited for glimpses of the cards to come, which they did. Except this time it was much more confused. Each card seemed to have at least two or three cards superimposed on top of each other. For the life of me I just couldn't seem to see a dominant card in any position. It was just some fuzzy superposition of alternatives.

At this point I could have given up; obviously just statistics at play; 14,400 to 1 is not anything to get too excited about. Then I thought, well why did I get glimpses of the cards in the first place; why did the impressions degrade in quality/certainty? It suddenly entered my head that instead of trying to "see" the card's face values, I might be able to "measure" the *weight* of each card instead. Ace is the lightest, five is the heaviest; oh and yes these cards *float*, I said to myself. Each card will float (in my mind's eye) to a position according to its weight. The ace will float higher than the five, with each intermediate card being appropriately positioned.

So again I shuffled and laid out the cards. This time I waited passively for the cards to obey my command to move themselves to their appropriate positions. After a short time I caught a glimpse of the cards all positioned at different heights. Another couple of glimpses verified the ordering. I turned over the cards; ace—hit, two—hit, three—hit, four—hit, five. OK, something is going on here; let me do this again. So again I shuffled and laid out the cards; again I waited passively for the cards to obey my command. After a short time a glimpse comes; but what are those faint ghostly impressions (outside the dominant impression) I see? A couple more glimpses verify the dominant positions. I turn over the cards, this time slightly less confident than before: ace—hit, two—hit, three—hit, four—hit, five. I think I see some pattern, some features emerging here. OK, I'm up to 207,360,000 to 1. Again I go for the third run: shuffle and lay; some moments; a glimpse; fuzzy superimposed impressions; a couple more glimpses to verify the impossibility.

So where to go next? Maybe I'll just instruct each card to flip over (in my mind's eye) and fly away in ascending face value order. Shuffle and lay; a few moments; flip, flip, flip, flip, flip; (verification) flip, flip, flip, flip, flip: ace—hit, two— hit, three—hit, four—hit, five. Again: Shuffle and lay; a few moments; flip-flip, flip-flip, flip, flip-flip, flip; (verification) flip, flip, flip-flip, flip-flip, flip-flip; semi-confidently chosen ordering: ace—hit, two—hit, three—hit, four—hit, five. Dare I try again?: Shuffle and lay; a few moments. STOP FLYING AWAY!!!

"Enough," I thought, I'm stiff, I'm drained; I've got the message. At this point, the odds (2,985,984,000,000 to 1) of my success by chance seemed less important than the possible importance of the features I seemed to have unearthed.

So there we have the experiment that gave birth to my focus into the area of contextual differentiation. Craft each context or question individually (subjectively) and create a unique targeting structure each time. Use that target structure for the purpose it was created, *then dispose of it.* It is of no further use to you—an ice sculpture made for a wedding party, it has melted, it has lost its form.

Much too good to be true? Just some unnamed computer geek pulling the legs of gullible parapsychologists? I emphatically make no evidentiary claims about this anecdotal, unsubstantiated report. But note well that whatever one's doubts or skepticism, he points out simply: "How easy, though, to test." This is precisely why the gates of orthodox science must be opened a little wider to invite in just such experiments, conducted in a spirit of curiosity and without prejudice.

I hope the link with an evolutionary analysis of the function of psi is apparent. Our hunter-gatherer ancestors did not sit around the fire in the depths of the Pleistocene era, entertaining themselves on shivery winter nights playing cards and trying to outsmart one another by using the psi honed in finding food and shelter. It's enough that if psi is a reality, they surely did hone it in that daily, imperative fashion. Imaginative precognitive exercises with cards can perhaps reactivate and recapture that skill, at least until our sophisticated minds grow bored and start yapping for extra novelty.

What's more, if psi functions as an adjunct to the rest of our faculties, we might expect it to conscript, or be conscripted by, the regular feature detectors and formal structure generators of the mind. Ever since Noam Chomsky introduced generative grammar fifty years ago, we have known that speech and much of the intellectual ordering expressed through speech is made possible by a profound internal and inherited grammar, a sort of machinery of mind that sets up a menu in infancy from which our experience selects certain options. Thus, we learn to speak the same language as those around us, and with the same accents. But those languages, however superficially various and mutually incomprehensible, are all formed on the same armature. Cognitive psychologists and neuroscientists have found numerous instances of preset forms and patterns into which our perceptions are poured, like jelly into a mold. Might psi make use of these formal constraints, this hunger for order and shapeliness? The evidence suggests that it does, often to our disadvantage, alas, because preconceptions and bias can twist the faint information from elsewhere in time and space, forcing it into line with our routine expectations. But there might be a way of tapping into this deep-seated appetite for pattern.

☙☙

Recall the ball-drawing experiments conducted by Suitbert Ertel and his students, which attempt to make use of our very ancient and intuitive capacities to see, reach, grasp, and capture things in the world: a ripe fruit to eat, a buzzing mosquito to swat, a parent's hand, a stone to fling, a child's body to lift carefully in both hands and hold protectively against your chest. Dr. Ertel has recently devised a brilliantly intuitive and startling method of tapping into psi, using that same fundamental and incredibly simple technique. This time his participants reach into two separate bags, without looking, simultaneously withdrawing a numbered ball in left and right hands, attempting nothing more than to achieve a certain rather abstract shapeliness created by the succession of paired numbers. Here is a version of Stanford's "goal-oriented" psychic prowess that, if found to be repeatable by others, casts a remarkable light on the interaction between psi—at least, in high-scoring subjects—and our aesthetic impulses.

To date, these experiments have been conducted long-distance between Germany and the Ukraine, overseen only by the participants themselves, without controls or scrutiny. Ertel justifies his faith in participant honesty by pointing to their significant positive results from tests conducted both in person and in computer trials of extrasensory perception that can't be tampered with (unless the experimenter is in on the deception). "Independent Freiburg experimenters and I myself tested [the two major participants] in 2003 and we confirmed their psi power," Ertel told me. "If a person's I.Q. is 150, obtained by an intelligence test under control, one need not be concerned that the person's exceptionally good homework might be due to fraud." The test itself is eerie but as simple as the British computer specialist's card precognition described above. The task: the participant will draw from separate bags one numbered table tennis ball with each hand. Each bag contains 50 of these balls, 10 each numbered one through five. Just in case the participant drops or jumbles the balls in a particular draw, those in the two bags are colored differently. The participant continues this drawing process for 180 trials, constituting a single run. (In several cases, participants completed 181 or 182 trials instead of 180, missing the moment at which to stop.) Before each trial the participants shake the bags, and after each trial the balls are put back into the bags.

Now we get to the interesting part. No efforts were to be made to pick a particular number with either hand. There were no traditional targets. Instead, as pairs of numbers were drawn, they were first listed on two separate sheets, one for each hand, and then used as coordinates to specify a cell in a five by five matrix. That is, each double draw specified a single entry in the appropriate cell on a square with five columns and five rows, with 1 through 5 across (rows) representing the numbered ball drawn from the bag on the right, and 1 through 5 vertically (columns) representing the numbered ball drawn from the bag on the left. So if you blindly plucked out ball 3 with your left hand, and also ball 3 with your right, a heavy dot is placed right in the center of the matrix. If you drew out ball 1 with your left hand, and ball 5 with your right, the dot goes instead in the top extreme right-hand cell. Remember, there was no instruction to select any particular numbers. Ertel's instruction was just to elicit "by mental activity such as hope, desire, concentration, as far as possible, *non-random accumulations which eventually form a pattern that should look orderly and pleasant.* The participants are encouraged to rely on their individual feeling of orderliness and pleasantness, no external norm need to be considered" (my emphasis).

The results speak for themselves, although it is not entirely clear what they are saying, other than that psi appears in these cases to promote not only an uneven spread of drawn numbers but a powerful aesthetic component as well, lending itself to eye-catching pattern making. By chance alone, the dots should fall utterly at random within the matrix, spattering the five by five array of cells like sparse raindrops falling with utterly no destination in mind. Instead, Ertel's participants, including three generations from a single family in Kiev—seventy-two-year-old Galina, her daughter Tina, and Tina's son Vanya, and a student named Amelie—have generated remarkably symmetrical or striking patterns, apparently entirely unconscious of the process involved. Perhaps it reveals a kind of domestic synchronicity. Dr. Ertel summarized the effect: "The left hand knows what the right hand does, in real time."

Again, one must caution that these results were obtained at home, without the beady eye of an independent observer overseeing the activity. Ertel promises that for the next stage of experimentation, "codes other than numbers, indicating the coordinates, should be used, 25 different codes, say letters. Code e in the left bag might represent row 1, code k in the right bag might represent column 1, etc. The test person should be unaware of the codes' key. The test person should only want to produce a nonrandom, orderly and nice-looking picture. What will happen?" Nobody knows—yet. Ertel emphasizes repeatedly that his current results are just intermediate or preliminary reports on exploratory tests of a novel method, yet to be corroborated. But if these astonishing portraits of anti-chance can be replicated, we'll have fresh insight into what psi is doing, and perhaps what it is used for. It is imposing order upon the disorderly aspects of the cosmos, despite its somewhat tricksterish nature. In the next chapter, that will prove to be a possible key to the very nature of the phenomenon.

The clearest and most revealingly aesthetic way to display the results of these gifted Ukrainians is to show the score for each cell as a deviation of its vote from the average number expected purely by chance. Galina's first and third tests are shown below. Her most remarkable effort to date, made in August 2006 without immediate feedback and recorded by her daughter, is shown in the second figure. From the viewpoint of those simply trying to assess a psi-

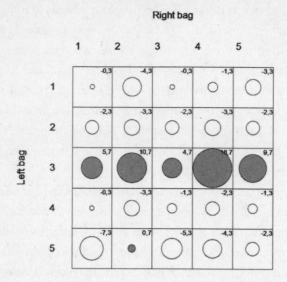

Figure 7-1. Galina's first test, with feedback. N. trials = 182.

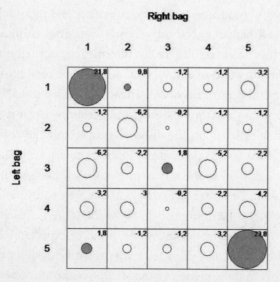

Figure 7-2. Galina's third test, without feedback. N. trials = 180.

induced anomaly, the key is the notably increased variance of dot accumulations.

Since there were 180 two-handed trials in this run, each trial nominating one of twenty-five cells, we expect by chance to see 7.2 votes in each cell (plus or minus a standard deviation or two). If one particular cell is so disfavored that it receives only a single vote, its score will be registered as -6.2. If another is the particular favorite, picking up 22 votes, its score will be displayed as +14.8. If nothing but random chances of work, we would expect most of the cells to contain quite small black and white circles. Instead, Galina's psi-directed draws produced this striking anomaly (figure 7-2). A Chi-square test estimates a chance probability of one in a million.

Nor was this single, apparently artistic accident a trick of the eye found after trawling through a huge hidden stock of obviously random results. Suitbert Ertel obtained dot distributions from additional participants with obvious, even though less strong, deviations from chance. He selected as participants people who had obtained high scores in the ball test under standard conditions.

Figure 7-3 shows Tina's result, using 2,880 draws, and the associated chance probability is less than 10^{-15}. Figure 7-4 was produced by Amelie, also with 2,880 draws, which is not significant overall, but the Chi-square estimate for the rows is one in a billion. If these images are less visually arresting than Galina's, still it is obvious that something very, very nonrandom caused these distributions. Was it simple deceit? Ertel is persuaded otherwise, but we will have to wait and see how readily, and to what extent, these amazing results can be repeated under controlled conditions.

Oh, and in case you were wondering what a purely random result would look like, figure 7-5 on page 253 shows one of Ertel's computer simulations of 2,880 double-handed draws. Not in the race.

☙❧

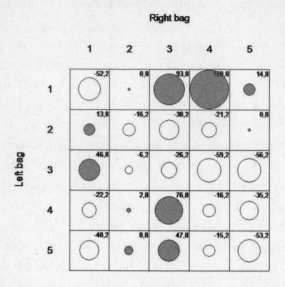

Figure 7-3. Tina's result with 2,880 trials.

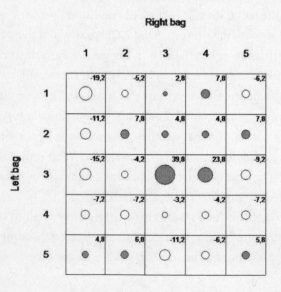

Figure 7-4. Amelie's result with 2,880 trials.

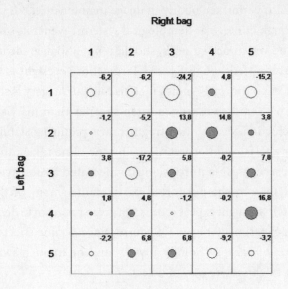

Figure 7-5. Computer simulation comparison of 2,880 double-handed draws.

It is one of the more charmless gifts of the alleged "intelligent designer" that once we hit age sixty or so, all kinds of things start to go rapidly and increasingly wrong with our bodies, not least with our sensory equipment and, to some extent, our cognitive capacity as well. Specialists in mental imagery such as Stephen Kosslyn and Itiel E. Dror, and Fergus I. M. Craik and Erik Dirkx, have studied this deterioration. Paola Palladino and Rossana De Beni report that older adults "appear to have problems in mental imagery which seem to influence memory, such as in the recall of concrete words. However, the effects of imagery instruction on old participants' memory are quite inconsistent. . . . Significantly, older participants failed in image activation and image rotation, and also performed poorly in image maintenance."[7] If psi is expressed through our regular neurological systems, even if its source is elsewhere and very much stranger, should it not also degrade in an analogous fashion? (Although Russell Targ, whose eyesight is very poor—10 percent of normal—notes that his mental imagery is sharp and clear, so his remote viewing is often sharper than his vision, although psychic images are often fragmentary, requiring stabilization and later analytical interpretation.) By the same token, to the extent that observed psi is unconsciously mediated by the *experimenter* rather than by the purported participants (many of them young students), we might expect results in a given program to decline slowly as each experimenter ages. Of course, since decline effects are part and parcel of parapsychology anyway, and on a much swifter timescale, this might not be easy to detect.

But then some experimenters and participants who are still getting excellent results are not exactly spring chickens—Suitbert Ertel was born in 1932, and Galina's results at age seventy-two are even more impressive than those of her daughter and grandson. Dr. Jim Carpenter told me: "I have a database of about 700 ganzfeld trials from several labs, ages 15 to 75 for percipients. Age and ESP scores (ratings of likeness between ganzfeld imagery and targets) had a zero correlation. Also, when the ESP scores are squared (to make them express degree of extremity irrespective of direction) these scores also had a zero correlation with age" (personal communication of an unreported analysis, September 15, 2006). Well, if there isn't any psi decline with age (aside from ordinary attendant deficits in the regular senses), might this be regarded as evidence for a nonmaterial component to the human psyche? Then again, who is to say that "nonmaterial" implies resistance to decay or fragmentation?

Stephan Schwartz offers an observation drawn from decades as an explorer of anomalies, although I am far from persuaded by his implicit theory of psi:

The physiologic aspect may be an aspect, but not a determinant variable. I presented a paper some years ago at an anthropology of consciousness conference that showed the scores of Alan Vaughan, Hella Hammid, Judith Orloff, and maybe six or eight other remote viewers across years, and many experiments. They actually got better over time. From my perspective, the same would be said of such people as Eileen Garrett, or Stefan Ossowiecki. I think this occurs because being able to open to nonlocal awareness is a function of focus. Paradoxically, I think it can be likened to becoming like a little child because, like a child there is no judgment, just an openness. It's not surprising that this is an empirically developed observation made again and again across many cultures, places, and times.

It might be that any physiological deficits that result from aging are offset by the growth in mature poise, judgment, or skill. Schwartz adds:

Focus is necessary to suspend intellectual activity, judgment, and analysis, learning instead to report the purest possible sense impressions and sense of knowingness. Across the decades of getting to know people who are good at this skillset, it has become clear to me, at least, that one either acquires focus through emphasizing neurotic behavior, or through some kind of discipline, meditative, martial, religious, artistic, or mathematical. As long as the task is not deadening, and sometimes even when it is, if there is enough novelty (as Rhine and Pratt showed), engaged individuals get better. Decline occurs when the task becomes detached, mechanical, and rote, not just when we get older. (Personal communication, September 15, 2006)

Thus, experience in field and lab corroborate the British computer expert's inventive approach on the best way to tap psi: novelty, playfulness, and relaxed openness to what might feel like memories of events you've never experienced. A kind of tender persuasion of otherwise random events in the direction sought by a mind hungry for pattern, shapeliness, and agreeable surprise. Why would we have these abilities? The evolutionary answer seems to be: to make our way more successfully in the world, and perhaps to bear our genetic treasures the more securely with us down the stream of time. But the question returns: How can this apparent magic fit in with the universe as it is revealed in deep scientific equations and subtle experiments? What kinds of adequate, indeed robust, theories might make sense of these disturbing, these affronting anomalies?

8

INTO THE WOODS

I may have a unique position in psi research: I spent the first 12 years of my physics career in experimental nuclear physics. Thus I lived, breathed, and was totally immersed in quantum things. So while there are some good theory people around (for example, Brian Josephson) I think I am the only one that has a gut-feel based on years of experiments with mixed states—what's now called "entanglement" by the New Age and some parapsychologists. In my view, sadly none of the latter know much about the concept. Just because one can find a few (and I mean a *very* few) physicists who say what the new agers like does not make them correct. One example: To most physicists there is no difficulty whatsoever with regard to the experimental results proving nonlocality. Einstein was dead wrong with his too often quoted "spooky action at a distance." In short, I don't think quantum mechanics will have anything to do with the final understanding of psi.

—Edwin C. May (personal communication to the author)

Not everyone loves quantum theories of consciousness, let alone quantum theories of psi. There seems no doubt that psi requires an explanation going well beyond the restricted laws of physics as they apply for the world of ordinary experience. At the same time, it seems that the uncomfortable laws of the quantum realm do share some of the properties of paranormal phenomena. Psi appears to be fairly convincingly nonlocal in character. But such weird quantum attributes are not supposed to show up in our macro-level world. Most scientists agree that entangled connections between distant states simply don't, and can't, exist except in very rare and contrived circumstances. Then there's the vexed question of whether entanglement is even capable of explaining paranormal knowledge or action, given that formal quantum mechanics insists that no information or energy can be exchanged via entanglement. For all of that, some notable parapsychologists do insist that quantum nonlocality and entanglement are probably more than metaphors, that they provide deep insights into the nature of psi.

Dean Radin is one of these parapsychologists. In his 2006 book, *Entangled Minds: Extrasensory Experiences in a Quantum Reality*, Radin insists that "the key to understanding psi resides in the possibility that the universe is a quantum object. That means everything, the whole shebang, is a holistic, coherent unity. That includes our bodies and minds. . . . I must make clear here that entanglement as presently observed in elementary atomic systems is not sufficient to explain psi. But entanglement that emerges within biological organisms, what we might call bioentanglement, will be a giant step in the right direction." Stephan Schwartz, in his 2007 book, *Opening to the Infinite: The Art and Science of Remote Viewing*, uses the expression "nonlocal awareness" for a number of experiences he regards as sharing a common root: psi, mystical or transcendental experiences, aesthetic bliss, genuine creativity. "What we think of as precognition seems much more likely to be nonlocal awareness of probabilities not yet actualized by the final acts of intention that bring them into what we think of as 'reality.'" But Schwartz remains cautious: "The words

nonlocal awareness imply that this is a quantum phenomenon, perhaps a form of entanglement. This may be the case but, I want to be
clear, this has not yet been proven."

The problem with regarding mind and psychic capacities as
derived directly from quantum weirdness is made clear in an exasper-
ated comment by Ed May: "We should be very careful how we throw
terms like entanglement about without putting in proper caveats
about the obvious and well-verified QM conditions to maintain
those interesting states." Dean Radin's brain on earth, at 300° Kelvin,
he points out, simply can't be entangled with a rock on Mars at
200°K. The kinds of preparations required in laboratories to maintain decoherence are finicky, extreme, and not very long-lasting. Consider, for example, a recent paper boasting "robust entanglement," by
H. Häffner, F. Schmidt-Kaler, and their colleagues, in Innsbruck, Austria, who commented: "It is common belief among physicists that
entangled states of quantum systems lose their coherence rather
quickly. The reason is that any interaction with the environment
which distinguishes between the entangled sub-systems collapses the
quantum state. Here we investigate entangled states of two trapped
Ca+ ions and observe robust entanglement lasting for more than 20
seconds."[1] Nearly half a minute, two calcium ions, using three powerful laser pulses! This is wonderful news for anyone who hopes to
build a quantum computer, but it has no bearing whatsoever on the
suggestion that remote-viewing the rings of Jupiter is due to an
entanglement state between a brain here and some observation point
hundreds of millions of kilometers away in space, let alone a brain
here and another (or even the same brain) three days in the future.
How could that entanglement state be created? How could it be sustained between two such different environments?

Admittedly, an even more recent paper published in *Physical
Review Letters* by a team of physicists in the United Kingdom, Austria,
and Portugal in February 2006 claimed that "photons in ordinary

laser light can be quantum mechanically entangled with the vibrations of a macroscopic mirror, no matter how hot the mirror is. The result is unexpected because hot objects are usually thought of being classical. The finding suggests that macroscopic entanglement is not as difficult to create as previously believed and could have implications for making room-temperature quantum computers in the future."[2] Again, though, this kind of quantum entanglement would have to be established with considerable effort by experimental scientists, because at room temperature, macro-systems have long ago lost their quantum coherence and just don't exhibit wavelike behavior. It's not the kind of thing that happens automatically, or while you're dreaming or in a remote viewing daze. May and his colleagues put it crisply: "This is not to say that a comprehensive model of anomalous mental phenomena may not eventually require quantum mechanics as part of its explanation, but it is currently premature to consider such models as more than interesting speculation. The burden of proof is on the theorist to show why systems that are normally considered classical (e.g., a human brain) are indeed quantum mechanical; that is, what are the experimental consequences of a quantum mechanical system over a classical one?"[3]

How could the brain of an operator sitting in front of one of PEAR's random number generators get entangled with the inner workings of the generator? Assuming it does get entangled, how could that entanglement possibly last long enough to complete a run of attempts to bias the outcome alternately in two different directions? I've seen quite breathtaking examples of hand waving in this regard, starting from the proposition that *everything* was once entangled with everything else in the moment of the big bang (a highly doubtful assertion anyway), proceeding to the supposed but absurd consequence that such entanglement persists to this day. Radin, in an eloquent summary of his recent book's argument, says:

The entangled minds concept proposes that the physical medium we live in is already and always entangled. Entanglement persists after the big bang because atoms (and photons) are constantly interacting. I am entangled with suns in the vicinity of Zeti Reticuli because I interact with photons that originate there. Entanglement is nonlocal temporally as well as spatially, suggesting that everything is entangled, even if it isn't yet (it will be!).

Classical smoothing due to decoherence provides a convenient approximation that allows us to use simple Newtonian mechanics to build bridges, but it doesn't mean that the entanglement magically vanishes. Thus, if it's always there, then at some "deep" level below what our senses tell us, we live in a holistic medium. All we need do then is ask, what would human experience be like in such a medium? The answer is the entire list of psi phenomena.

In sum, nothing needs to be created or sustained in this model. It's all quantum, all the time. (Personal communication, August 2005)

I remain unpersuaded, for the reasons enunciated by May, but I recommend that the interested reader consult Radin's book for a different and well-defended opinion.

So if signals of the traditional kind don't seem to be a good explanation for either anomalous cognition (telepathy or clairvoyance) or anomalous perturbation (psychokinesis), and quantum entanglement seems rather far-fetched, are we forced to accept some kind of quantum observer explanation? Do we really create reality by *observing* into existence—by bringing about state vector collapse—one out of many superposed possibilities? Something along those lines is asserted by Jeffrey Schwartz, Henry Stapp, and Mario Beauregard in a 2004 paper published by the *Philosophical Transactions of the Royal Society,* hardly a fly-by-night psychic news sheet. Veteran physicist and theorist Stapp was at the Lawrence Berkeley National Laboratory, University of California. Schwartz and Beauregard are neuroscientists. Their argument is not conventional, however. They reject the assumption that mechanistic brain functions explain consciousness, noting that in contemporary physics "the consciousness of human agents enters into the structure of empirical phenomena." Granted, once a particular action has been chosen, the consequences follow relentlessly according to the deterministic character of Schrödinger's equation, but "the choices made by human agents must be treated as freely chosen input variables." Consciousness, therefore, must arise far outside the boundaries of strict determinism—and once this is admitted, the same process that couples conscious decisions to the meat of the brain and body can be conscripted to explain so-called paranormal phenomena. Their argument continues with a deployment of equations, analysis of synaptic function, and so on, but it is painfully familiar to any philosopher, as are its built-in flaws.

In what sense are these input variables "freely chosen"? Whether they arise from computations within the brain, or from some spooky ghost hovering above the machine, it's clear that they can't be *random*, in the quantum sense of utterly undetermined. What kind of freedom would that be? In the 1970s there was a brief craze for a novel called *The Dice Man*, by Luke Rhinehart, in which the increasingly crazed narrator governs the course of his life by writing down a blend of sensible and lunatic options, then throwing a die to see which one he would follow for the day. Not surprisingly, this provides unnerving and hilarious fun for the reader, but not for the narrator. This doesn't mean that some form of dualism is not the case, because, after all, the intuitions of most of humanity insist that there is a spiritual, nonmaterial aspect to personhood that endures beyond death. But even if there is a ghost in the living machine, it would be a very unsatisfactory ghost if its choices were governed by random chance.

I am not saying that Schwartz and his colleagues make that claim. But if a suitably deterministic and reliable ghost is the best we can get in order to retain our sense of judgment, character, and responsibility, then why not stop with a suitably deterministic brain, without introducing this invisible component?

Nonetheless, once this ghost is proposed, it's true that it needs something like psi in order to couple with its body, and it is not obvious why such a power must stop at the boundaries of the skin. Does it follow, though, that the existence of psi necessarily implies the existence of a nonmaterial spooky source of consciousness? Psi might provide the basis for an extended physics that could find room for postmortem survival, and who knows what all else—free-roaming spiritual entities that never knew the joys and restrictions of the flesh; loitering souls afraid to take the next step "into the light," as many mediums insist; hapless or vengeful souls or demons eager to infest the living; and so on. Dualism probably seemed like a good idea at the time, but the road to hell (literally) is paved with good intentions. As William of Ockham put it: *Don't multiply hypothetical entities unnecessarily.*

William's principle forms part of the basis of an alternative analysis proposed in 2005 by Dr. Harald Walach and Dr. Stefan Schmidt, two German specialists in complementary and alternative medicine. Much of the recent exploratory theoretical work in anomalous phenomena has been conducted by German or Dutch researchers, some of whom have several doctorates apiece. (Walach has one in clinical psychology and another in philosophy and history of science; Schmidt's is a psychology doctorate.) After presenting brief summaries of the copious evidence supporting the reality of psi phenomena, and while admitting the deficiencies of some of that evidence, they argued persuasively that a powerful reason for skepticism from mainstream scientists is the baggage attending most claims of the paranormal. Certainly this is something I've noticed myself, sometimes to my alarm, among theorists of psi.

> The reluctance of modern scientists to accept such anomalistic phenomena as real is not only that they do not fit into a prevailing world picture, but also that they seem to come with a theoretical framework in tow—namely, ontological dualism or parallelism, which is considered outdated. We make a plea not to overlook the data for fear of buying an unwanted theory.

That data, that accumulating bundle of anomalies, points to some central failure in the way reality is represented by orthodox science, which as every sophisticated onlooker now recognizes is itself supported by a barrage of ideological preconceptions, special interests, and political pressures, in addition to its immense storehouse of facts, hypotheses, and powerful theories. At some point, if the evidence in favor of psychic phenomena breaks through the resistance of those who refuse to look at it, a paradigm change (a term now sadly contaminated by black holists) will surely be required to contain it.

What kind of fresh approach do Walach and Schmidt expect? Nonlocality, they suggest, is undoubtedly a key. Suppose that "the basic stuff of the universe is neither matter nor consciousness after all, but some transcendent element which is basic to matter and mind at the same time and in the same way. . . . Such an approach would not necessarily demand exotic kinds of energy or unknown types of fields," although that could turn out to be the case. Quantum theory would then not be the bottom line, but just an example of "a more generic type of nonlocality." Easy enough to say, but what kind of deeper-than-deep model are we talking about here? String theory? Branes? No. Luckily, Walach and Schmidt have one to hand, the so-called weak quantum theory, developed under the auspices of physicist Dr. Harald Atmanspacher at the Institut für Grenzgebiete der Psychologie und Psychohygiene.

Exploring that new modification of an already excruciatingly difficult topic goes beyond the bounds of this book. In essence, weak quantum theory

> employs an algebraic axiomatic approach which uses only the most basic and general formal structures characteristic of quantum mechanics, while omitting more specific definitions. For instance, it does not provide a definition of addition and subtraction and thus cannot make precise probabilistic predictions. However, it defines multiplicative operations, both commutative and non-commutative, and hence accounts for the most central feature of quantum mechanics, namely its non-commutativity, which is the algebraic expression of complementarity. . . . A natural consequence of complementarity is entanglement between sub-elements of a system.

Don't worry, I don't really understand that either. The bottom line is that "WQT is *one* example of a theoretical structure that is able to account for nonlocal processes such as those observed in [psi phenomena], without making additional assumptions, postulating strange entities, unknown fields or occasional violations of natural laws." No signals are required in bringing about this nonlocal connection between distant states. All that's needed, Walach and Schmidt claim, is a *correlation* between the two. Can such a correlation account for precognition, telepathy, even psychokinesis of the kind of observed in poltergeist infestations (usually regarded these days as spontaneous psychokinesis produced by someone under stress, and constituting a plea for help)? Not, so far as I can see, if those phenomena are to extend into the realm of repeated success at remote viewing on the scale found during Star Gate.

Yet another alternative model of psi advanced by a senior German anomalies specialist is the so-called model of pragmatic information (MPI) developed in the last couple of decades by Dr. Walter von Lucadou, originally a physicist specializing in chemical reactions induced by streams of charged atoms. Twenty years ago Lucadou was conferred with a second doctorate, summa cum laude, this time in psychology at the University of Berlin, with a thesis on "Experimental Investigations of a Possible Influence of Observers on Stochastic Quantum-Physical Systems." This sounds at first like a variation on Evan Harris Walker's theory, but MPI is derived from systems theory, a field of study more advanced in Germany than elsewhere, perhaps as a consequence of the immense prestige of Jürgen Habermas, a philosopher and sociologist of the pragmatics of communicative reason and action.

In systems that are "organizationally closed," the pragmatic aspect refers to the *meaning*, for the system, of some quantity of information. For example, a participant in a psi experiment will become enclosed, as a system, with a random number generator, inducing meaning-rich correlations of a nonlocal kind. Or so says Lucadou. I have to admit that I haven't the faintest idea how this cashes out in terms of the underlying physics. In any case, Lucadou deduces several profound similarities between the components of such pragmatically linked systems and mathematical structures of quantum theory. For example, as noted earlier, it's key to quantum mechanics that certain properties, certain observables, are what are called "conjugate variables." The link between these complementary observables limits the degree to which one or the other can be measured exactly. If you know precisely *where* a particle is, you can't determine its exact *momentum*. The more exactly you know *when* something happens, the less exactly you know how much *energy* is involved.

Similarly, Lucadou concludes, the novelty of a certain psychic effect can only increase as the capacity to confirm its detection diminishes. *Autonomy* (the freedom to choose a particular kind of test) and *reliability* of psi manifesting itself are involved in this kind of mutual trade-off. The first few times a striking phenomenon is investigated, its impact can be considerable. But once these observations are used as pilot studies, setting the rules for future confirmation studies, the search for reliability damages the effect sought. This might sound like a magical incantation rather than an exercise in theoretical physics, but Lucadou has applied his calculations to a number of reported psychic phenomena, not least some vigorous poltergeists (recurrent spontaneous psychokinesis, or RSPK) and discerned a predictable two-part pattern, the second part comprising four testable components that seem to embody his prediction ("Predictions of the Model of Pragmatic Information about RSPK," Walter V. Lucadou & Frauke Zaradhnik, 2004).

The first part of the pattern is that the structure and function of poltergeist activity will be revealed as a sort of externalized psychosomatic symptom of an unrecognized inner malaise. This much, of course, is traditional. Where it starts to get interesting is in the four-stage unfolding predicted by Lucadou: (1) a *surprise* phase, which comes out of the blue and can contain horrendously shocking and unexpected events, the sort of thing seen in horror movies like, well, *Poltergeist* (abrupt movement of heavy objects, even levitation and perhaps teleportation of objects); (2) a *displacement* phase, where the living human psychokinetic focus or source is identified and perhaps accused of trickery designed to attract attention and sympathy; (3) a phase of *decline* in the effects, which might well be attended by deliberate fraud, meant to extend the excitement and attention; and finally, (4) a phase of social *suppression,* the general, somewhat conspiratorial, decision that nothing really happened after all, certainly nothing worth getting anxious or excited about.

And indeed, in a way, Lucadou—as a counselor for people in distress—doesn't really care one way or the other whether true psi was manifested, or if fraud replaced its initial appearance. His systems-theoretical approach has located disruption, and now he can help remedy the underlying problem that the poltergeist effect was meant, however unconsciously, to unmask. Lucadou's general position, in respect of all psi phenomena and not just RSPK, is summarized thus:

> In contrast to the usual observational theories (OT), the Model of Pragmatic Information (MPI) does not start at the description level of quantum theory but at a very general system theoretical level. It roughly says that psi-phenomena are nonlocal correlations in psycho-physical systems instead of signals or forces. Such nonlocal correlations, however, limit the psi-effects due to the conditions of the psycho-physical system, which [are] mainly described by the "meaning" of the situation, i.e. pragmatic information. . . . [MPI] does not limit the effect size of effects even though the model does not assume PK strong violations of physical laws and moreover assumes that PK is not a real force but a non-local correlation . . . the structure and duration of an RSPK-case (its history) and the number of hierarchical levels, which are necessary to describe the organizationally closed system, determine the size of the emerging psi-effects.

This analysis might sound sensible enough, if rather baffling in its appropriation of certain aspects of quantum theory that you might consider suitable only to the realm of the very small, but Lucadou draws from it a rather startling implication. As noted, the moment a paranormal phenomenon falls under the microscope of scientists and becomes routine—"bureaucratized," as it were—it must start to fade away or slip sideways into some other unexpected manifestation. And this is precisely what most anomalies experimenters have found to be the case, to their annoyance and frustration.

Probably the most extreme instance is the international Freiburg-Giessen-PEAR consortium's recent attempt on a truly enormous scale to replicate the PEAR database. The overall results showed a reversion to the mean, effectively a null outcome that critics were quick to interpret as evidence that the previous two decades of work at Princeton must have been profoundly flawed, if not indeed a result of deception. Careful study of the fine-grain detail of those results showed, however (according to the PEAR analysts), that several significant and meaningful anomalies persisted in the background of those ostensibly failed data. This kind of belated discovery cuts no ice, naturally, with skeptics, and perhaps it shouldn't. Except that a model such as Lucadou's predicts precisely this kind of displaced outcome. Not a reversion to chance, which is the skeptical assumption, but an irruption somewhere else, like the bulge that pops up in one part of a sausage balloon when you squeeze it in another.

What's more, Lucadou's model does seem to imply a general decline effect, not abrupt but slow and steady, as any once-interesting psi protocol is more widely adopted and repeated. As we've seen, that's precisely what Bierman found—although he also found hints that after a long no-show interval, psi effects could rebound and recover. What, then, of outstanding exceptions to these declines, such as the reported continuing effectiveness of Star Gate–trained remote viewers like Joe McMoneagle? Simple; it's exactly the security-classified nature of that original research that protected it from becoming routinized. Usually the remote viewers were not told their target and were not even provided with feedback about whether they had been accurate or not. Well, maybe. That still doesn't explain how McMoneagle is able to continue scoring so exceptionally well two decades after leaving the covert program. True, his targets are usually free response and conducted quite often in the glare of television cameras, which perhaps acts to deinstitutionalize his performance.

So can anything be done to ensure success of psi experiments? Lucadou suggests:

> According to the Model of Pragmatic Information, the components of pragmatic information in poltergeist cases are alternatively determined by two goals. The first . . . is the internal goal of the organizationally closed system to produce an effect in society. The second one is the external goal of society to prepare the organizationally closed system. The internal goal affects the combination of novelty and confirmation, while the external preparation of the system has an effect on the autonomy and reliability of the system. Only autonomous systems can produce novelty, a necessary component for a phenomenon to be evaluated as anomalous or psychical. To preserve the autonomy of the system, however, one should not prepare it in such a way that everything has been determined. The system "can only behave as it pleases" as long as one does not observe it with great care.

ᘒᕲ

Perhaps this reminds you of the image of the trickster from anthropology. Like Loki in Norse myth, this mischievous and sometimes cruel figure from Native American folklore intervenes whimsically in human and natural affairs, sometimes bending the rules, sometimes apparently breaking them. Coyote is probably the most famous of these pranksters. Brer Rabbit and Bugs Bunny are two more. A postmodern and rather baggy evaluation of parapsychology by George Hansen attributes many of the woes, uncertainties, and startling delights of psi phenomena to what he calls the trickster, although I have to admit that I still can't work out precisely what level of reality he attributes to this figure. It seems elusively interesting, but exactly what is being proposed here? There's such a vast and murky discourse eddying along behind this buzzword. One could mistake what's being said for some cautionary remarks on chaos and emergent complexity (mathematically defined), but in fact that line of thought would tend toward the contrary conclusion. Chaos math has shown us that many apparently random, turbulent, or evasive phenomena are driven by a few simple "grammatical" rules deep within a system. That would be order within disorder, not the reverse.

Is there a hint of an intrusion into our physics-describable world of intentional beings of some other order than the one known, or even knowable, by physics and chemistry—the kinds of entities previously called gods, demons, revenants, or spirits? These would then be deliberate tricksters, playful minds that sport with us from, say, a higher-dimensional realm. Or perhaps the trickster story is just anthropomorphic: actually we impose this naive narrative on events that make sense only if perceived from that putatively higher-dimensional standpoint. It would seem that the subject matter of anomalies research doesn't just jitter around stochastically but actively *eludes* the researchers' gaze and grasp. It resembles one or more intentional minds—now showing itself, now ducking for cover. If we were to take the claims of a palpable trickster like Uri Geller, say, at face value, we'd need to assume that these mysterious volitional entities, or vast cosmic currents, or whatever they are, have selected a conjuror who frequently fakes his performances—sometimes rather clumsily, from the couple I've witnessed—only to scandalize us with genuine bursts of paranormal activity. (Stanley Krippner noted as far back as 1975: "Geller was not above using sleight of hand when he was nervous, tired, or when his powers faded.") Yet tricksters shun most laboratories, or come and go whimsically.

Such an account is perhaps not strictly meaningless, but it strikes me as an atavistic retreat from the Enlightenment's insights into world and mind. It looks like the kind of category mistake made when our hard-wired feature detectors are set off by a chance cloud formation, turning the cloud into a face. We are selected by evolution with what cognitive psychologists now call a "theory of mind" that helps us empathize with and interpret our kin and our foes, and it all too readily misfires when we look at anything more complicated than simple variants on regular order. Which is why anomalies researchers need to be extremely careful about their interpretations. Do they start to feel that there are Minds there, providing paranormal glimpses and nudges, or that *we* spill over into a larger realm where our powers are larger but, being beyond our conscious control, not especially useful to us in crossing the road without being run down, or in earning a living? Is this going further than simply saying that there are factors we don't understand (everyone knows that, after all), rather than proposing that the universe is intelligent or alive (to put it coarsely)? Or that it harbors mysterious volitional entities that make the life of a parapsychologist a misery? Let us keep gods, demons, and tricksters at bay as the hypothesis of absolutely last resort. Yet if there is one or more intentional anomalous modality of cognition or action (call the putative ensemble "psi" for short, while bearing in mind that it might be a mixed package), you would certainly expect previous cultures to have known this and developed social structures to contain and manipulate it. They did so with sex, nutrition, cleanliness, and architecture, so why not with psi? If so, what would such containment look like? Well, witchcraft, shamanism, homeopathic medicine, and other magical or religious practices seem like good bets. And historically it was just those reports, those memories now suppressed wherever possible by the gatekeepers of science, that gave rise repeatedly to the suspicion that humans do have access to abilities that are not accepted by reductionist and mechanistic sciences.

It's not as if human history for eight thousand years chugged along in a pragmatic way, never imagining that people might be able to coerce others by sorcery, send messages at a distance, see the future in dreams, levitate, or contact the dead until one day in North Carolina a former plant physiologist got the weird idea of tabulating some guessing games with cards, and found significant deviations, and a light went on over his head. ESP! I'll be durned! *Yet that* is *how electronics was born,* more or less. And neutrino astronomy, and proteomics. And Gödel's theorem, and Schrödinger's equation, and all of those other unexpected umlauted innovations. So to the unprejudiced eye, it might seem that parapsychology should be to witchcraft and spiritualism and tricksterism as physics is to alchemy. The alchemists were mostly dead wrong, and too bad, but they were the start of a long, painful path toward comparatively cleanly designed experiments and operationalizable theories. Should the Hubble technicians and Mars probe rocket scientists check with the ancient alchemical texts? (No, but some more work on arithmetic might sometimes be helpful.)

Ah, but this case makes the cardinal error of conflating *physics,* a nonintentional science, with a discipline where human purposes and emotions are integrally entangled. You can't treat people like cogs; they have to be treated . . . well, semiotically, to use a buzz term. As bearers and initiators, that is, of *meaning.* So does this imply that we're back to looking at witchcraft for clues to useful rituals, for purity and danger markers? It wouldn't do a lot of harm, and might do some good, if a few more people took the trouble to immerse themselves in these ancient practices and bodies of lore. But those doctrines are obviously also encrusted with millennia of bad guesses, ignorant folk theories, bone-deep category mistakes, projection, and unfalsifiable claptrap. So let us listen patiently but with a keen measure of skepticism, understanding that these doctrines and practices have evolved around powerful memetic attractors that did our ancestors a power of good in their small neolithic bands, and are still able to suck our brains into the soupy maelstrom of charismatic power trips or effective decorticalization. Then we should head off to the lab again and try to find out what it is that makes Joe McMoneagle do it right, when he does, and what gets in the way when he's wrong. The absence of a pentacle and chicken gizzards is probably not a factor. (Just guessing.)

At the same time, we mustn't allow ourselves to forget that the most powerful paranormal phenomena ever claimed have been reported by schizophrenics and other mentally disordered people, especially while they were in states where they were unfriendly to the rational order. More than one untreated schizophrenic has claimed to experience himself or herself as being a divinity able to control the entire cosmos, something perhaps few mystics would have the hubris to announce. On a minor scale, a typical claim of the mad is the ability to change traffic lights at will—a small trick, but one that would evade the magic powers of many traditional mystics, I suspect. I am *not* asserting an identity between mystics and the mentally deranged. I *am* saying that crazy people, and people locked inside cult rote, frequently make wildly preposterous "paranormal" claims. We don't take them seriously (by and large, if only because their claims are mutually inconsistent).

Suppose a man tells a doctor "he had a vision where loathsome creatures seemed to crawl on the walls of his room. Then a man appeared who claimed to be God. This apparition said that [the visionary] was to be the one who would communicate the teachings of the unseen realm to the people of the world. He would be the means by which God would further reveal Himself to the world." Eighteenth-century mystic Emanuel Swedenborg, who had a very famous veridical precognition of a fire in distant Stockholm menacing his home, told his doctor just that. That worthy's negative reaction to the report seems to me entirely understandable, not a devious reaction, necessitated by his role as cultural liminality gatekeeper, to a very reasonable claim.[4]

So what is it about the claims of parapsychologists that make them more believable? I assume that it is the witness of their accumulating evidence and theoretical apparatus, however incomplete, by people who are *not* mystics or cultists.

It might be that to put the matter this way is *already* to adopt one side of a legitimately multisided debate. Perhaps if the universe *is* infested by a trickster aspect, there is no way any explicit practical and theorized program can produce consistent outcomes and insights, beyond an acknowledgement of Murphy's or Sod's Law ("If something can go wrong, it will"). But the history of science over the last four hundred years has been a series of encapsulations, of informational compressions. It would seem very surprising if psi were the only phenomenon at odds with this consilient vector. Can we resolve the problem, this *misleading* appearance of a trickster dancing in the midst of the anomalies? In his book *Be Careful What You Pray For, You Just Might Get It*, a leading proponent of paranormal healing, Dr. Larry Dossey, mentioned what he dubbed "the Legion Principle—the displacement of one person's problem onto another." Perhaps the universe might slam an unprotected healer or investigator with a kind of backlash, or evade the investigation with a deft Lucadou side step. There are more alarming possibilities.

❧

As you may recall from chapter 2, Dr. Elisabeth Targ, Russell Targ's daughter, studied glioblastoma multiforme after her study on AIDS patients led her in that direction, and subsequently contracted the rare disease herself. I don't know if she had any direct contact with patients suffering from glioblastoma, but we do know that at least some carcinomas are viral. Others are due to specific environmental insult. Might she have contracted the disease via contact with a patient, or that patient's environment? My wife's father was a radiologist, who died comparatively young from a rare cancer of the parotid that's unusually common among radiologists, presumably elicited by stray radiation. But I don't think he got cancer because the diseases he diagnosed became displaced on to him by magical symmetry; it's much simpler, if equally tragic, to blame radiation leakage. If chance coincidence fails to account for Dr. Targ's cancer and we wish to consider nonstandard explanations, several come quickly to mind.

It seems Elisabeth Targ was raised from childhood to take paranormal events for granted, including transtemporal effects. She expected to die at forty-two and was only off by a couple of weeks. If precognition informed her of this fate or destiny, it might well have let her know well in advance (unconsciously) that she would perish from glioblastoma multiforme. Because of this advance warning, she might have been alerted for that disease among her patients and developed a special interest in it and them. No need, if this were so, to posit any malign forces in nature or supernature. Psi precognition would explain her interest, and without imposing any toxic temporal loop that I can see. Knowing in advance that one of her cells would randomly mutate in that way probably would not change the probability of its occurring. By contrast, knowing you were destined to lose a tooth to decay might forestall that very outcome by sending you off in time for dental treatment.

Another possibility is that some kind of morphic resonance, a sort of psychic prion, can warp the genome of one of your cells into a fatal configuration if you get too involved with someone who is already infected. This is (barely) testable statistically. Do more oncologists get cancer than other doctors? Of course such specialists are also exposed to viruses from patients, and might have chosen their discipline precisely because of paranormal foreknowledge of their own fated cancer and a desperate unconscious wish to solve its threat (or a perverse wish to embrace it, as a psychoanalyst might claim).

Here's a still more disturbing possibility along those lines: might an interest in a given disease, coupled with an unconscious will to self-harm (or other motive), allow the mind to use selective PK to mutate a cell into the relevant precancerous state? This is a possibility I intensely dislike, since it's a "blame the patient" suggestion. Still, it could be true, and might be the other side of the coin from selective PK *correcting* of a mutated codon, which thereby (perhaps) heals a patient's tumor or prevents its growing in the first place. Note that infliction (which involves only a single cell) might be far easier to evoke than to cure (which requires the repair of many cells in a tumor mass), so perhaps hostile sorcery is more common than blessed healing. Arguing against this general line of thought is growing medical knowledge showing that, for example, while Papuans who ate human brain matter thought they got deadly kuru because of sorcerous curses (and killed other villagers in fearful punishment for their surmised wicked psychic emissions), it turned out to be due to a slow prionic infection from their gruesome diet.

The strangely hypostatized "trickster theory" doesn't seem to help here. Inversion and then deconstruction of dominant binary oppositions is a function of human analytic procedures, not of the brute universe. It's something *we* choose to *do*, not something the cosmos *does to us*. (Unless, of course, the cosmos *is* animate and intentional, full of malign or tricky spooks, demonic malevolence, the influence of UFO Grey implants, and so forth, but I rather hope we have moved on past that kind of superstition.) The high status of psi practitioners in hunter-gatherer societies is not a result of their affinity with a literal trickster, but of their culture's (historically understandable) lack of any real grasp of physics, cosmology, infection, or mental illness, and their gap-filling adoption of ideas of reference, animist projection, superstitious fears, and co-optation of schizophrenic and other weird utterances. The same dynamic applies to the adoption of such ideas by deranged and deranging cults. Accepting that psi is a reality does not imply that it supports the beliefs of the mad. Had electrons been conceptualized as angels rushing through conducting wires, it might have been decided that using their passage instrumentally to toast bread and run a desktop computer is blasphemous and spiritually ruinous. Even allowing that high-level (or high-flown) spiritual claims are not of this sort, any metaphysical starting point that grounds psi in God, a primordial mind, or a trickster principle seems likely to yield an approach to the topic that is extremely at odds with one derived from two-cheers-for-reductionism materialist science, suitably amplified where necessary by quantum weirdness.

Can coupling between intention and matter or energy cause warping of what we might call a *probability field?* Probability space, on this model, would "bend" in the presence of some psi element of consciousness, rather as physical spacetime is curved in the presence of mass, producing a local, temporary change in the probability density. But to suppose that wishes directly "bend" probability space, such that the external universe contorts its trajectories to fall into line, strikes me as an entertaining metaphor but a dubious theoretical model. Suppose we wish to account for the presence of so many skyscrapers in big cities. One could surmise that for any given city, there is a probability space that maps the range of structures it might comprise. So how do we explain New York? Is it best accounted for by saying that the original bustling city chanced to contain a number of psychically powerful "probability field changers" of a megalomaniacal cast of mind, whose wishes warped the town planning probability space in such a way that buildings in Manhattan simply got taller?

In a sense, yes, but what actually happened is that a multitude of social contributions (including the presence of megalomaniacs driven by the desire to put their stamp on the landscape, plus restricted territory around the central business district, plus the utility of having such a compact district) coincided with technological discoveries or inventions—steel girder construction, the elevator, the telephone, and others—allowing very tall buildings to be built. The "probability space" is a *description* of the way these contingent factors interact. It is true that one can, in a meta-way, modify the empirical outcome of city design by deliberately engineering people's attitudes, or putting money into elevator research, but varieties of quantum theory instantiate probability fields as a kind of *thing*. Walker's theory can be read that way. But to me (a nonphysicist, admittedly) it makes sense only as a metaphor. "Probability spaces" are contingent on physical spaces, mental tools derived from our attempt to understand physical spacetime and its stochastic regularities. In that sense, what is being described is real, rather as esprit de corps is real. But most physicists seem to regard a construct like "probability field" as a tool for calculation, not a description of the ontology of the universe. Consider Schrödinger's own view. Here's a passage from Walter Moore's biography, *Schrödinger: Life and Thought* (Cambridge UP, 1989):

> In June 1946, Schrödinger had written to Einstein: "God knows, I am no friend of the probability theory. I hated it from the first moment when our dear friend Max Born gave it birth. For it could be seen how easy and simple it made everything, in principle, everything ironed out and the true problems concealed. . . ." At about this time, however, Schrödinger wrote two papers on "The Foundations of the Theory of Probability." He had been reading a little book on this subject by Pius Servien, which suggested that an unsatisfactory basis for the concept of probability might cause serious problems for physics. Servien thought that all attempts to define "probability" beg the question. "They

seize the word and torture it, but can extract from it nothing but itself." He thought that "probability" may be a word like "beauty" rather than a word like "potential." Schrödinger therefore tried to set forth an approach to the concept from certain first principles. . . . Schrödinger adopted [a subjective interpretation]: ". . . Given the sum total of our knowledge, the numerical probability p of an event is to be a real number by the indication of which we try . . . to set up a quantitative measure of the strength of our conjecture or anticipation, founded on the said knowledge, that the event comes true." (436)

A psi account explicitly using a probability field model has been advanced by physicist Dr. Claude Swanson in a large book, the first of a promised series, that might be described by his admirers as admirably syncretic and by others (I am one) as an indiscriminate hodge-podge. *The Synchronized Universe* (2006) analyzes the somewhat fallible process of precognition via an ingenious parallel drawn from dynamical systems theory (the formerly fashionable chaos theory).

Our passage through time, Swanson suggests, somewhat resembles the path of a ball rolling and jogging down heavily rutted, furrowed hillsides and valleys. Looking down from above, from the vantage of some additional dimension of time, we see that the ball is generally channeled by a winding, dry creek bed. Now and then it splits into two or three rivulets. The place where the creek bed divides is flattened, so the energy required to jump from one rivulet to another is low. The slightest breeze might divert a rolling ball from one potential pathway to another. Up above, we can see easily that if the ball had taken path A, one set of futures would open up before it, while taking path B leads it in a completely different direction, with distinctive opportunities. It's even possible for one tributary to loop back and join another. Such an analogy helps explain why remote viewers can generate with considerable confidence a consensus picture of a future event until a certain date has elapsed, and then find themselves creating a completely different picture. The subject of this precognitive clairvoyance has reached a branch point and taken a less likely path.

Whether or not this image is appropriate to precognition, such a potential surface or landscape might serve as an analogy for probability fields that present opportunities but not destinies.

In recent decades, another fascinating, controversial, and vexing alternative way of looking at the problem of psi has been advanced in some detail by May, Spottiswoode, Utts, and others, and perhaps bears some resemblance to Lucadou's MPI. Their decision augmentation theory, or DAT, is a sort of mirror image of Schmidt's theory of quantum psi, which construed all kinds of psychic phenomena as variations of PK. Instead, DAT sees it all as due to opportunistic selection by the participant, or perhaps by the experimenter, of brief random fluctuations in the state being observed at moments when the fluctuations happen by chance to be favorably biased to the desired outcome. Such selection is guided by unconscious precognition. This might seem to replace one unknown absurdity with another, but since precognition is already recognized as a reality by anomalies researchers, even though it is not yet explained, why not press it into service to account for as much as possible in the psychic menagerie?

Here's how it works: Every day, every moment, the world is offering us choices. We reach decisions using a combination of distilled past experience and new information and act upon that blend of data. Because we are not simple, programmed machines, we don't act entirely in a canned and predictable fashion—although quite a lot of the time we do, which is why people are able to make fairly reliable judgments about our character and how we are likely to behave. The usual assumption is that the data on which we base our decisions, and our actions, is restricted to whatever enters our senses or is provided as a kind of prerecorded stereotype from our genetic inheritance. But if psi is a reality, we presumably obtain at least some additional but incomplete information that should weigh upon our choices, even if that information comes at us under the horizon of our own conscious awareness. But suppose that precognitive knowledge allows a sort of PMIR black box inside our heads (or inside our souls, for that matter) to foresee the cascade of alternatives we're about to face. Then we might be able to make a choice that skews our path in the direction of the desired outcome, whatever it happens to be at the moment.

Imagine you wish to meet a friend in a busy crowd. There is a certain strong likelihood that she will be walking down a particular street sometime between five and six o'clock. You are in a hurry and can't afford just to wait there on the street hoping to catch a glimpse of her in the scurrying crowd. How might psi help you in this task? According to the traditional view, you might be able to place a message in your friend's mind, directing her toward a rendezvous. Or your desire to see her at a certain place might be detected by her own psi, unconsciously, of course, directing her steps toward you. A somewhat unpleasant variant might suggest that you use a form of psychokinesis to modify her brain state, persuading her to meet you at this arbitrary intersection. The DAT explanation is very different. Precognition allows you to know in advance when she will be at a certain place at a certain time, and an unconscious calculation permits you to set off on your own stroll at just the right moment to make the rendezvous. Is this a reasonable notion? It seems to require not only the capacity to foresee the future, but to make an enormously complex calculation and then act on it unerringly. What's more, presumably your friend is also modeling your future actions and either choosing to evade your strategy, or collaborating in it.

As usual with such conjectures, it is far easier to test them in a simple toy universe. Borrowing from recent parapsychology protocols, we can imagine an activated PK-testing machine with its internal randomizer running, flipping states thousands of times a second. Once you press the button that registers the start of a run of trials, your job will be to generate either a HI or LO bias in the machine's output, or just to leave it running purely at random for a baseline (BL) vote. Suppose producing a HI requires 101 or more of the first 200 binary flips to be 1 rather than 0 after you press the button. Similarly, LO will be registered if fewer than half of the flips are 1. Traditionally, we'd suppose that a psi-successful HI vote would be achieved by *manipulating* the randomizer, *forcing* it into conformity with our will. But remember that the stream of random numbers is rushing onward whether or not we push the button. Some sequences of 200 bits, by chance alone, will have an excess of 1's, while others will have an excess of 0's. According to DAT, your mind has the capacity to reach into the future and detect the moment when a sequence is diverging from chance in the direction that you wish to see, while you wait, finger poised over the button, for just the right initiating instant. The details will be slightly messy; it takes measurable time for the brain to make a choice, send information through the spine, down the arm, and into the muscles and sinews. It takes measurable time to depress the button, and for the signal to pass into the machine and register the beginning of your vote. We have to assume that this delay is also taken into PMIR account—perhaps by training, perhaps simply from a lifetime of experience—since it is assumed that every decision we make is modified slightly by this kind of augmentation. The upshot is that when the stream of too many or too few 1's is coming your way, you get the chance to register that fact and to make a decision based on it.

Notice an interesting implication. Without some kind of psychic force modifying the randomizer output, the best you can do is choose the optimal moment in a completely random sequence. But suppose once you start registering your guesses, you are obliged to continue making an additional guess every thirty seconds, following the HI, BL, LO schedule. After several minutes the random sequence will be, by definition, entirely uncorrelated with anything you would like to see. On the other hand, suppose you get to choose the starting moment of *every* trial with a separate push of a button. In that case, there's nothing to stop you from repeatedly optimizing which part of the flowing random sequence is selected—assuming that precognition is incredibly micro-accurate, that you have psychic access to the streams of flip states inside the machine from near future (or even past, if the stream is prerecorded but unseen), and that you can press the button at precisely the right time to grab that sequence and not some other one less to your taste.

It seems that it would take superhuman coordination to use DAT effectively, which might be why most experiments with random number generators are only marginally more successful than chance; such acuity can't be sustained. Indeed, May and his colleagues insist, "The domain in which DAT is applicable is when experimental outcomes are in a statistical regime (i.e., a few standard deviations from chance). In other words, could the measured effect occur under the null hypothesis? This is not a sharp-edged requirement, but DAT becomes less apropos the more a single measurement deviates from mean chance expectation (MCE). We would not invoke DAT, for example, as an explanation of levitation if one found the authors hovering near the ceiling!" (1995). What's more, according to this model it doesn't make any difference whether the source of the statistical variation is classical or quantum in nature. All that's needed is precognitive access to that future state.

I have to say I remain troubled by this idea. Surely, you think, there are a hundred ways to prove that it can't be true? Why not make your starting point for the experimental run dependent on something like, oh, the barometric pressure recorded in the weather report precisely seven days hence in the *New York Times?* May and his colleagues are there before us. What made us choose *seven* days as one of the parameters? Might it not just as easily have been five days or six days? Why that newspaper, rather than the *Washington Post,* the *Times* of London, or the *Times of India?* Why barometric pressure rather than temperature? All of these available choices, we are given to understand, are winnowed in advance by the DAT precognitive engine, whatever that proves to be. This model—which is not a formal *theory,* as its proponents are careful to point out, but rather a way of compressing and coordinating the available laboratory data or "phenomenology"—might still seem like special pleading. The crunchy part comes when its advocates show, as they claim to have done, that almost all extant databases for anomalous perturbation or cognition of random binary states produce precisely the curve predicted by DAT, rather than any of those generated by alternative hypotheses. That has to give a scientist pause. (It has not impressed those at PEAR, however, who insist that their data refute the DAT model, whether evaluated by probabilistic or Bayesian criteria.[5])

Can DAT even explain Joe McMoneagle's ability to find missing people halfway around the world in Japan? What is being selected and augmented? The random jittering of parts of his brain, as it waits in repose, probing the workings of its own imaginative unconscious? That's possible, perhaps, but something in me repudiates the notion as a somewhat desperate bid to contain the paradoxes and absurdities of psi. Can this information-not-force model explain *any* aspects of anomalous perturbation, supposing that PK really exists? It's hard to see how. But the road toward an effective and comprehensive theory of psi remains long, shadowed, and arduous, and if anyone cares to set out on the tough journey down that road, I suspect they'll be wise to start with the pioneering maps already sketched in some detail by Ed May and his colleagues.

9

TOMORROW'S PSI

The reasons why Prigogine believes that he can dispense with the observer is mathematical, and I cannot pretend to understand them: "We arrived at a realistic interpretation of quantum theory because the transition from wave functions to ensembles can now be understood as the result of Poincaré resonances without the mysterious intervention of an 'observer' or the introduction of other uncontrollable assumptions". . . . He argues that it is possible, and indeed necessary, to combine the indeterminism of quantum mechanics with the view that probability is an objective quality of the universe, and not of our statements about it. I cannot understand this.

—Michael Frayn, *The Human Touch:*
Our Part in the Creation of a Universe

What is the take-home message of the psychic mysteries we have examined in this book? Must we turn aside, finally, as baffled by psi as the brilliant novelist and playwright Michael Frayn, challenged in his lack of scientific expertise, confesses himself to be by the no less wonderful mysteries of advanced physics? That would be a counsel of despair. None of us can know everything, even with the benefits of remote viewing. What I hope to have shown in the course of our voyage is how early exploratory wanderings along the track of the paranormal, first emphasizing the qualitative, spiritual, *storyteller* side of the experience, led step by increasingly austere step to the brutally, boringly reductive, *bean-counter* procedures recommended by J. B. Rhine and his colleagues, and finally to a drastic break with those critically important but stultifying transitional protocols and back to the rich possibilities of immersion: ganzfeld, remote viewing, even the continued investigation of spontaneous poltergeist phenomena and perhaps global psychic resonance and an aesthetic dimension to psi.

It certainly seems clear to me that psi is established as a reality, even if an excruciatingly irritating one, unreliable, puckishly evasive. If it works for Joe McMoneagle today, there's no reason to suppose that it failed to influence the guesses of Rhine's subjects. If psi can create unexpected aesthetic patterns, if it can bring you the hair-raising sensation of somebody looking at you from seclusion, and if it can modify or perhaps predict the output of random number generators, it seems extremely likely that it is an evolved function of the human species, playing its contributory part in our survival and thriving. Perhaps it is involved in the very process of maintaining or even optimizing the human genome against the depredations of chance. But how are we to explain its apparently magical operation? Is it indeed a spooky function of quantum observation? Or do we, somehow, routinely cast our minds outward and forward via some unknown medium, testing the world for threats and opportunities? In the laboratory, do we adapt that capacity for presentiment, allowing us to select the optimum moment for action that will bring about the result we desire, however abstract, even absurd, that might be? (And surely there are few more absurd and abstract tasks than guessing a sequence of random binary numbers, or attempting to skew their mean value up or down, aren't there?)

❧

Is there any way to use the paranormal to investigate its own physics? That would be elegant. If paranormal apprehensions can plumb the future, even a provisional future, can psi tell us anything informative about the future of psi? Or does that involve a toxic loop, lifting yourself by your psychic bootstraps? If Stephan Schwartz and his untrained remote viewing students were able to predict with uncanny accuracy the scene of Saddam Hussein's capture, we might hope for expert psychics to fetch back a treasure trove of insights from some future time—generations hence, or just a decade or two—when the secrets of psi are finally unmasked and put to use. Stephan rejects the question as wrongheaded: "I think psi is normal human functioning, in no way anomalous or paranormal. The future of psi is to disappear, even as we recognize that nonlocality of consciousness is essential to understanding how the world works. As parapsychology is withering in the U.S., and becoming a minor non-threatening aspect of psychology in Europe, nonlocality is a growing consideration in experiments ranging from biology, to medicine, to physics" (personal communication, September 26, 2006).

Both Schwartz and Joseph McMoneagle have ventured into the future and past in numerous dedicated remote viewing sessions, and attempted to sketch out histories of coming decades and centuries. In 1998 McMoneagle presented a sort of *pointillist* narrative from the late 1990s through to the year 3000, with side trips back to the time of Jesus Christ and the apparent garden tending of pre-humankind by some kind of cloudlike energy being. Schwartz is more modest in his goal, attempting a consensus RV portrait of life in 2050. The difficulty with both sets of purported time-trawled data is common to all remote viewing exercises, but it's especially disruptive when there's no possibility of immediate feedback: the observer's mind is obliged to fill in the many gaps, to translate the confusing sketches and blurts of news from tomorrow. How would a remote viewer in the time of Leonardo da Vinci have understood a vision of a helicopter, or a desktop computer? Even if Leonardo actually did get some of his inventions from the future (which is not impossible, given the reality of psi), he had to render them in terms that were familiar to himself and the technological climate available to him. As, indeed, he did.

This process of opportunistic invention is called *confabulation.* A hypnotized subject in a bare room given a suggestion can confabulate it into a full-blown hallucination. Elephants decked in gold and strawberry yogurt march before her bewildered gaze; fish with lacy wings cluster in the corners of the ceiling; she rises from the chair in a bubble of light and plucks the silver apples of the moon. All of this seems perfectly compelling. It is entirely unconscious in origin, as dreaming is, and fills in the gaps that brute reality has left empty. From the available literature on remote viewing and telepathy, it seems very likely that much of what's reported is sheer confabulation, sometimes verging on insanity. And a large part of the protocols developed under the Star Gate program and elsewhere are devoted to an end run around those almost inevitable contributions from fancy that fill in for us the details and sometimes the grand architecture of what we learn partially and inadequately by psi.

With that caution in mind, let's consider the future as explored by Joe McMoneagle. In his rather vexing 1998 book, *The Ultimate Time Machine,* prediction after prediction is scattered across the page in no particular time order, loosely gathered under headings such as Economics, Environment, Social, and Technology. After September 2001, McMoneagle noted, there would be a stock market slowdown. As indeed there was. That might seem a pretty impressive prediction, since it was made several years in advance. But in another sense, since we can't help reading this prediction as relating specifically to the destruction of the World Trade Center, it's also rather . . . well, *cold.* Let me add at once that this comment is not meant as a reflection on Joe McMoneagle's sensibilities—just that the report is oddly detached, given the epoch-marking tragedy we now retrospectively interpret it to be. How can an RVer pick up the stock market chart curve and miss nearly three thousand falling murdered humans? I put this question to Joe; his answer is sobering, but I'm not entirely sure that it's completely satisfying. "You mean as compared to 80 million deaths under Mao, or 4 million deaths in Cambodia during the late sixties and early seventies?" That is, a psychic is obliged for self-protection to insulate himself from the brutality and horrors running like sores across the planet or the planet's history. For all that, these Grand Guignol megadeath horrors do not seem to me directly comparable to 9/11 (but perhaps that's because I'm not a practiced RVer). Had those millions of deaths all occurred in a single hour or day—say in a nuclear attack—I *would* expect the emotion to register on someone precognitively tracking the economy of China or Cambodia. For example, it seems psychologically implausible to imagine someone in 1935 reaching a decade into the future to check the state of the Japanese economy and reporting simply, "In the first half of the coming decade, heavy industry will pick up markedly, but will receive a setback in orders after August 1945." This would be strictly correct, but absurdly elliptical. And there's the additional factor of *personal relevance:* clearly the 9/11 events have had a shocking and persistent impact on almost all Americans—indeed most Westerners—whereas perhaps the communist death camps in the East are not so vividly in our hearts.

We run again into the difficulty of *reliability.* Presumably this is the main reason why Star Gate was closed down, despite its tally of remarkable successes. Could government and military authorities be expected to draw upon precognitions when often they are so very much less than accurate? For example, what are we to make of the confident prediction in *The Ultimate Time Machine* that "The current pope will die late in the year of 1999. The new pope will be Italian (but not a native Italian), and take the name of Pius XIII" (169)? In fact, Karol Józef Wojtyła, Pope John Paul II, died in April 2005. His successor was not an Italian but a German, although he had lived for many years in Vatican City. Does this imply that in an alternative universe these events came to pass? Or that the death of the Polish pope was subsequently postponed, in our world, by divine decree? Or just that there's a certain flexibility in the timing, so this prediction will still apply to some papal succession, but in 2019, say, or 2999? It's easier to assume, I think, that McMoneagle drew this extrapolation from his memory, his knowledge of the world and current affairs, mixed with unconscious processes of imagination. The same might be said of his forecast that between 2002 and 2005 the Metaphysical Church of Science, or something along those lines, would be brought into existence to merge scientific knowledge and spiritual thinking. Its symbol, he says, is an infinity symbol superimposed on a circle. To my eyes, this looks rather like a scholarly version of the media's typical UFO Grey. There's no sign yet of this new faith, unless it's the Raelian UFO cult. By 2050, we're told, it will be doing quite well. More plausibly, Catholic priests will be permitted to marry by 2020.

McMoneagle provides a scatter of predictions in no particular order. Climate change will have a major impact on food production until 2070. Two large-scale boom-and-busts will be seen in 2020 and 2050 (dates McMoneagle finds slightly dubious because they are round figures and because he can't locate the reasons for them). NASA will be operating ground-to-space aircraft by 2075, which sounds rather delayed even for that bloated, sluggish agency. Major volcanic eruptions will occur in the Hawaiian Islands between 2014 and 2023, in Washington State between 2026 and 2030, and near Mexico City between 2029 and 2031. By 2075, houses will be smaller, electronically controlled by voice, and their walls will display holographic images of the outside world, as in Ray Bradbury's visions from the 1940s. By 2008, men and women will be wearing an androgynous "utilitarian" clothing style, rather like the jumpsuits of *Star Trek*. (As I write, that's not much more than a year away, and the fashion dictators of Paris and Italy don't appear to have any such ideas bubbling away on the catwalks, nor are such garments being sold in Wal-Mart. This looks to me precisely like a confabulation.) Formal contact with UFO aliens traveling in what amounts to time machines will occur between 2075 and 2077 (217), or alternatively will be delayed until the middle of the twenty-second century, due to our primitive behavior up to now (275). Either way, the aliens will surely be disappointed by the terrible wars that persist thereafter.

Let's jump right ahead to the year 3000. It's a green, environmentally stable earth, rich in diverse biomes, its inland cities of circular design, with white but stubby skyscrapers and outlying residential zones connected by underground transportation. The global population (perhaps two hundred thousand per city) seems to be markedly restricted, perhaps as low as one billion, made easier by implanted contraceptives so that pregnancy is always a choice. Power is obtained from a new device with a singularity at its core—a small black hole? A white hole? A stress that taps zero-point energy? After a ravaging religious war, a sort of ecumenical intelligent designer faith seems to have emerged, shared in some sense by profoundly modified versions of Buddhism, Christianity, Islam, and Hinduism. What they seem to have in common is a notable role for prophets, perhaps a spiritual version of today's psi practitioners. "One of the basic tenets . . . is that there is a single power within the universe that has caused all matter, both inanimate and adamant, to exist. . . . There is a strong indication that humankind has developed a much greater understanding of the role we play in the construct of reality . . . we now recognize . . . universal laws that govern our reality, our place within the physical, and how our actions affect the other species within it" (266–70).

In the year 3000 eugenic engineering has come and gone, although genomes are checked at conception and repaired if necessary. There are no taxes, banks, or wars. People get around by hover car or by a kind of teleportation limited only by the amount of power fed into it. You can use a phone-booth-sized portal to get around the city, although it takes something the size of a bus to travel between cities, larger "teletransporters" to get to the planets of the solar system, and singularity ships in orbit get you to "any galaxy desired." That's still a new technology in 3000, and apparently relativity problems make it hard to be sure you get when and where you're headed. Understandably— a teletransporter (let's hope they have a catchier name by then) "pulls space/time from one place to the next. Whatever happens to be contained in that space/time location is moved" (274).

This kind of projection of the wonders of the future astounded readers of *Astounding Science Fiction* seventy years ago, and perhaps still strikes many people today as far-fetched—what used to be called "that Buck Rogers stuff," but is known today by journalists as "that Star Trek stuff." It's a possible future, no doubt: a stable or nearly stationary utopia arising from several diabolical global wars. But look, this is meant to be *one thousand years* in the future! We're talking about a gap in history equal, at the very least, to the gap between the depths of the Dark Ages and the present day. But study those projections carefully, and they seem no more startling than the changes experienced in the rich First World countries in the last century.

For comparison, here are some of the projections offered by amateur remote viewers working with Stephan Schwartz. To date, they have made some four thousand consensus forays into the year 2050, and the project continues. Surprisingly, Schwartz is unfamiliar with the details of Joe McMoneagle's projection, but anticipates marked overlap, with 75 to 85 percent of both sets of material wholly or partially correct when evaluated by concept rather than by detail:

On the basis of my own experience in the future world, I believe that 2050—and, by extension, 3000—is a pretty good representation of what is coming. My problem with going too far out is that you can't really understand what they are saying—I can't understand what the energy boxes are that people talk about seeing less than fifty years into the future. There is a high degree of consensus that there has been an energy revolution, and that the source of energy is something magnetic that comes in a box. One goes to the store and buys a box suited to one's energy requirements, from flashlights to cars, or houses.

I find these projections remarkably staid, especially McMoneagle's, which strikes me as a blend of utopian H. G. Wells and Aldous Huxley in mystic mode. I can't even accept that the curve of technological change could stall for the next four decades at Schwartz's homely level. Go to the *store?* In an age where molecular manufacture (nanotechnology) should be commonplace, and almost free, as today's iPod downloads are? Imagine a team of remote viewers in 1956: "By the distant year 2000, gigantic mainframe computers will be found in every neighborhood. Ordinary people will be able to drive to them and buy complex calculations performed by specialized programmers to satisfy such needs as tax returns, business analyses on spreadsheets—and even games for the kids!" That looks so tame today; actually, it would have been spectacularly prescient. Yet see what's missing: computers you can carry in your hand, the global Internet, the global phone system, Google and other search engines, the human genome map, cloned animals, no Soviet Union—1956, remember, was a mere three years after the very structure of DNA had been first uncovered, and the height of the cold war. Throw your mind forward an additional forty years. Will you still be going to the store in 2050? What I myself anticipate, without benefit of psi, is exponentially accelerating technology leading to an opaque technological singularity caused by the construction of greater-than-human artificial intelligence, and all the cornucopia benefits and shocks that will follow.

Granted, I am drawing upon my own prejudices here. A decade ago, I discussed this proposition of madly onrushing technology in my nonfiction book *The Spike*, which tracks likely paths into the next half century and beyond on the basis of Gordon Moore's "law," the enduring observation that computer bang for the buck increases one thousandfold every decade. That is a jump of a million millionfold by 2050, when it is plausible that miniature, cheap home computers will hold processing power equivalent to the human brain—which is when things will *really* start to get strange fast. In *The Singularity Is Near*, the brilliant inventor Ray Kurzweil writes: "Most of the intelligence of our civilization will ultimately be nonbiological. By the end of this century it will be trillions of trillions of times more powerful than human intelligence." By contrast, McMoneagle predicts mildly that by 2050 "computer memories will experience a fifty-fold increase [over what was available in 1997]. . . . By the end of 2050, the average home computer will have 1,000 times the current processor power"—whereas, in reality, just ten years later, desktop computer random access memory (RAM) had soared from half a megabyte to a gigabyte or more, and processor power had already grown by a factor of more than one hundred. Stephan Schwartz, though, is not dismayed by the drastic implications of this Singularity challenge to current expectations:

> Stores haven't disappeared by 2050 but they are very different. There's little or no mass manufacture, mostly one orders up a model something, by designating the components that are then assembled, much like a Dell computer today. In the 2050 sessions, the usual description is that the energy revolution occurred some years earlier, and it took several decades for people—freed up with wireless global communications, and free energy—to adapt and move out of cities, creating a kind of new tribalization in which one

joins a community of like-minded individuals. Some are bohemian communes, some kibbutzes, some militaristic enclaves. You can already see this happening. Many years ago, the 2050s viewers accurately predicted the coming of HIV/AIDS—"a blood disease will cross over from primates in Africa and spread all over the world, killing millions"— and virtual reality. Their projections are not at all linear. Having been a professional futurist, paid well for my opinions, I can tell you the 2050 sessions are far more accurate, where they can be tested, than anything to come out from the professional futurist community.

This future resembles the best hippie dreams of the 1960s and 1970s, apparently lacking radical nanotechnology and major artificial intelligence (AI). The only justification I can foresee for these limitations would be draconian regulation, implemented to prevent rogue nano-factories outputting weapons, poisons, viruses, and so on. Which, alas, might well be imposed upon us. Of course, building both nano and AI might just turn out to be very, very difficult to do, or they might be shut down and interdicted before they are fairly begun by Homeland Security and its international equivalents. Still, I was startled by the lack in either Schwartz's or McMoneagle's predications of anything about major life-extension technologies. By the 2050s we should be seeing significant superlongevity treatments.

When Stephan began his project in 1978, he was deep into the geopolitics of Mutual Assured Destruction, or MAD, the strategic balance of terror between the West and the Soviet Union, about to get involved with citizen diplomacy between the USSR and the United States. His concern then was nuclear holocaust. He was taken aback, baffled, when the 2050s viewers said that one of the world's superpowers would disappear. "I couldn't imagine what that meant . . . Patently ridiculous." Yet a decade later he was in Moscow when the

White House was surrounded by tanks and Mikhail Gorbachev was arrested. He recalls staring down at Boris Yeltsin on a tank, seeing with a flash of insight that the 2050s RVers were correct after all. So tomorrow's world must be safer? "To which they responded—man, boy, woman, and girl—*No, actually, it gets more dangerous. There are lots of little nasty wars, and an army arises that has no state.* At the time, I had just come off the MIT/SecDef Panel on Innovation, Technology, and the Future, and was steeped in the thinking of the futurist community. I saw overpopulation and scarcity as the principle tropes of the future. I thought they were nuts. Who could mount an army but a state?"

Impressive, but still—no molecular cornucopias? Schwartz's answer shows the kinds of normalizing biases that can lock up such RV projections. "When I learned about nanotechnology in the '80s, I already had so much material I decided to stay consistent, so [I] don't know much about that. AI, which I did know about, is not something they ever talk about, although they do say that computers run most systems. They apparently don't see it as a major cultural issue." I find this as unlikely as supposing that the arrival on this planet of aliens from another star system would have no significant cultural impact. Similarly, I suspect that the viewers' projections for biological change reflect Stephan's own enthusiasm for alternative medicine and therapeutic psi healing: "Medicine is totally changed. Pharmaceutical medicine is mostly gone. Chronic diseases are handled pre-birth and, thanks to gene lining, heart disease, etc., is almost gone, except in the remaining pockets of non-technological societies." This is all entirely plausible, but probably insufficiently radical to be true. But Stephan argues that while the future "is not a linear extension of the present," he would wager that "the 2050s' vision of the future is going to be pretty close to what happens. We won't see all of it, but will see a lot. So far," he adds, "they have been scarily accurate."

Again, though, one has to ask: where is the psi in these futures? I am prepared to believe that parapsychological anomalies will continue to evade understanding for another forty or fifty years, as they have done for the last century or more. I'm even prepared to believe that they will be explained away definitively, their apparent mystery unmasked as a series of odd, overlapping side effects of mind and matter, like creativity, like laughter. I can't imagine for a moment that Singularity-scale transformations will not occur before the end of this century, and certainly before the end of the millennium in 3000—even if psychic revelations utterly rewrite our understanding of the world.

One of the artifacts we should possess well before 2050 is quantum computers, machines that access the multiple superposed states of quantum reality. Could these strange new tools of thought help account for psi? A computation is a sort of strictly organized thought. Might we, therefore, picture the universe not as the thought of God (the old idea) but as thinking a kind of God into existence, as process theology used to claim? "Some of the information processing the universe performs is indeed thought—human thought," Seth Lloyd has noted, but the vast majority "lies in the collision of atoms, in the slight motions of matter and light. . . . Such universal 'thoughts' are humble: they consist of elementary particles just minding their own business." As the universe expands and cools, though, Lloyd foresees life reaching "to encompass first stars, and galaxies, then clusters of galaxies, and eventually, it would take billions of years to have a single thought."

Here is a particularly disturbing idea: is it possible that the world we experience is neither the hard mass/energy reality it appears to be, nor a thought made manifest in the spiritualist tradition, but is instead a kind of computational simulation? To get a sense of the catastrophic possibilities that radical theory could hold in store, let us muse more deeply on the idea that the universe itself might be, in some sense, a computer.[1] Or, more precisely, a computer *program*, being run on a computer to which we could never have access— because we would be subroutines within the grand suite of programs. The late Heinz Pagels, as a skeptical high-energy physicist, put it this way in his splendid book *Perfect Symmetry: The Search for the Beginning of Time:* "An important feature of the quantum theory not shared by the earlier classical theory is that the information we obtain about the world by measurements depends on how we decide to obtain that information—the method of measurement. Quantum theory emphasizes information, its representation, and its transformation. Since computers also transform information, an interesting image for the quantum universe is that it is a giant computer—an information-processing system."

Ed Fredkin, one of the founders of the celebrated MIT computer science laboratory, has carried this notion to the extreme lengths of devising a new "digital physics," rather further down the track from the mere image or metaphor Pagels posed for our consideration. For Fredkin, the very laws of physics are nothing more than parameters controlling cosmic logic lattices or "cellular automata." Nor is he without notable admirers. "If anyone is going to come up with a new and fruitful way of looking at physics," the late Nobel Laureate Richard Feynman told science journalist Robert Wright in a book with the ironic title *Three Scientists and Their Gods,* "it's Fredkin." Although Pagels did not go quite as far as Ed Fredkin in his speculations, he took a huge first step. "In this metaphor of the universe as a cosmic computer," Pagels explained,

> the material things in the universe, the quantum particles, are the "hardware." The logical rules these particles obey, the laws of nature, are the "software." The universe as it evolves can be viewed as executing a "program" specified by the laws of nature although it is not a deterministic program like

those in digital computers. What the ultimate "output" of this cosmic computer will be remains to be determined. But we already know that its program has given rise to complex "subroutines" that we can identify with life. So complicated are these subroutines that they seem to take on a life of their own, independent of the cosmic computer.

This is breathtaking stuff. It is necessary to note that Heinz Pagels was a professor of theoretical physics at Rockefeller University, president and chief executive officer of the New York Academy of Sciences, and a member of the Science and Law committee of the New York Bar Association. (He tragically fell to his death, betrayed by a childhood polio weakness, while climbing Pyramid Peak near Aspen—and his book *The Cosmic Code* contains an eerie premonition of just such a death.) If life is in some authentic sense a kind of semiautonomous program running within a cosmic simulation, a rather Platonic conceit, might it not learn to interfere with the programming that constitutes that universe? Such self-programmed loops could produce astonishing discontinuities in the world that we (the subroutines) experience. One subroutine (or consciousness) might contrive to subvert the usual rules and directly access the knowledge normally available only to another data register (that is, a different person). It might even prove feasible to reprogram the physical conditions in one's simulated surroundings.

I am not necessarily putting this forward as a serious explanation for paranormal phenomena. My point is that heavyweight specialists working at the frontiers of established knowledge have not shrunk from advancing pictures of the universe that, unlike regular physics, do seem compatible with some otherwise inexplicable psi phenomena— indeed, in certain cases, predict their existence. If the world itself can be regarded as a computation, might it be computing a virtual reality, as in Hollywood's *Matrix* trilogy, or even an uncountable number of them, many inhabited by creatures that take the simulation for truth? Buddhism and Hinduism have ventured this way, as did the new age book *A Course in Miracles* channeled by research psychologist Dr. Helen Schucman (1909–1981). In transpersonal psychology, a version has been offered by Dr. Charles T. Tart: his "world simulation process" creates for each of us individually and collectively a biological-psychological virtual reality (BPVR).[2]

The idea has been around in science fiction for at least half a century, from Frederik Pohl's pivotal story "The Tunnel Under the World"[3] to its significant extension in Daniel F. Galouye's twice-filmed 1964 novel, *Simulacron-3*,[4] arguably derived from Plato's parable of the cave. Could psi be a result of occasional stochastic errors while running the code? Algorithmic simulation might conveniently explain how telepathy is possible, if every human being shares a common hidden communication protocol. It might help explain miracles or remarkable nonlocal awareness, which would require nothing more than the equivalent of a few key taps in the higher-order universe of our computational matrix. Strictly speaking, of course, this explains nothing about the origins, destiny, and purpose, if any, of the deepest substrate of reality, but it might explain a great deal about ours—if it is contingent in this way, if we are absurdly, whimsically elaborate fictions with consciousness.

Such cosmic vistas can be remote and terrifying, as well as awe-inspiring or even comical, but Lloyd's own computational cosmos has been laced with charming and sometimes deeply moving encounters with wonderful creators of simulations: with the brilliant Jorge Luis Borges, whose dense fictions first depicted in literary imagination a universe of infinitely forking paths; and with Lloyd's no less brilliant physics mentor, Heinz Pagels. "While he lived, Heinz programmed his own piece of the universe. The resulting computation unfolds in us and around us." Meanwhile, the cosmos continues its immense and star-blazing computation. Perhaps, since we're part of it, carrying forward our memories, that computation is not meaningless. If psi is an element in such a cosmic computation, or even the very ground of it, everything might finally be steeped in meaning. Must we, then, allow ourselves to be drawn toward the ancient suspicion, still the common opinion, that nonlocal consciousness—the spirit—persists beyond death, beyond the grave? That entities move among us, and beyond us, not of our own material kind but of some more enduring fabric— whatever that might mean?

The concept of a life for humans beyond our individual deaths is very old and very widespread—as old and universal, perhaps, as dreaming during sleep. Afterlife is often held by the nonreligious to be a consolatory fabrication devised out of grief and wishful thinking, an imagined realm where loved ones persist somehow beyond death as if they had traveled to a land beyond the hill or shore, a place where the evident injustices of mortal life are redeemed and set right, with punishment for the wicked and joyful rewards for the virtuous. Despite its evident gratifications, it seems to me that the wellspring of this idea is the real, confusing experience of half-remembered dreams.[5] When we sleep, our drowsing minds mingle memory and fancy, placing us or our viewpoint surrogate inside a kind of shifting, surreal virtual reality where time loses its implacable dominion, where the dead walk among us, and where strange chimeras are built from fragments of creatures, people, places, motivations, and feelings carried over from waking empirical life.

It is easy to see how such imagined worlds, vivid and more various than humdrum narrow reality, might have enthralled our ancient ancestors, undistracted by reading, movies, television, easy travel, or frequent visitors. Certainly we know that hunter-gatherers were given to punishing the living for slights or crimes experienced only in dream, in much the way diseases and accidents were widely blamed on sorcery and ill intent. But even if these are the sources of such widespread and poignant beliefs, are they necessarily untrue for that reason? Parapsychology suggests that intentions *might* act on others without any conventional medium of influence, and that thoughts can be intercepted even if unspoken. Is it possible that fantasies of life after life also offer us glimpses of a reality that scientific cultures dismiss because of their elusiveness and similarity to delusion and psychotic or protective self-deceit? The general background assumption of Enlightenment values, after all, has been that belief in afterlife will wither away as technology serves up utopia. Although it is true—despite rumors to the contrary—that more people now live more secure and comfortable lives than ever before in history, with life spans increasing in the privileged parts of the world, a suspicion grows that wealthy Westerners thrive at the expense of the rest, at the cost of a world rushing into greenhouse-effect and resource-depletion horror. Reincarnation might prove less tempting as a belief as that suspicion hardens. Steely disbelief, though, is a stoic virtue that is perhaps beyond the grasp of suffering people who must defer their hopes to a better life beyond the grave. Unless mediums and parapsychologists can demonstrate unequivocally that such a domain is real and attainable, its adherents will regard afterlife as something to be hoped for in private faith, rather than by watertight public evidence, and that posture is antithetical to the spirit of science, even paranormal science.

Classic vitalism, humankind's favored explanation for what sets life apart from dead matter, has been known for more than half a century to be mistaken. That scandalous fact was fixed at the heart of biology with the discovery in 1953, by James Watson and Francis Crick, of the simple chemical code that compiles unliving matter— molecules of carbon, hydrogen, nitrogen, phosphorus, and oxygen— into life. If any doubt lingered, it was quashed in July 2002, when a polio virus was literally assembled from brute components strung together in a laboratory according to its documented genome. While a virus is not strictly alive, since it needs a host to make its copies, there can be little doubt that an entire bacterium will soon be built from nothing more than a genomic recipe plus a handful of lifeless chemicals. But that, of course, is how life proliferates anyway, parent to child, in a lineage stretching back to the first simple chemical replicators on a barren world. In this sense, then, life requires no spirit breathed into its nostrils before it awakens from dust. The right kind of common elemental dust, guided by an evolved, conserved molecular recipe and energized by the sun's thermonuclear light, does the job of being alive all by itself.

Many people, perhaps most, still tremble at this fact, preferring to turn aside in denial. Surely such raw lessons from biology can tell us nothing important about life, especially human life, can they? Even a child knows the difference between a rock and a kitten, the one inert and changeless, the other squirming with joy and naughtiness, like the child herself. Yet the deepest mystery for an adult, the mystery withheld from the child's eyes as long as possible to protect her innocence, is that all warm flesh someday will chill and stiffen into nothing better than stone. Those sweet lips breathe no more nor speak to us; those poor, dead eyes stare without seeing, stones indeed. The heart we loved, which loved us in turn, or hated us, is stilled; spirit has left its dwelling. But has it gone anywhere—if it is indeed a thing rather than a process? Has it found another home? Is it nestling, even as we mourn, into unformed embryonic tissue that is fated to unfold in utero into a new person? Or does it sport (or suffer for sins of the flesh) in someplace untouchable by mortal hands, unreachable by living tread? Humankind has held tight to those assumptions, those terrified hopes, for a hundred thousand years.

Evidence has always seemed abundant, even aside from our desperate wish and consolatory belief that it be so. In dreams, our lost ones return to us, if we are lucky (or perhaps unlucky, should their message be vituperative). Analogy tells us that a flame blown out is gone, but only for a moment; a spark will rekindle its fire. A song dies on the lips, but is harbored in the heart, in the folds of the brain, until we call it forth effortlessly once again to move the air in patterned waves of beauty, melancholy, martial zeal. Must it not be like this with the human spirit? Is not death merely a temporary occlusion of the light, never truly lost if always somehow altered for a time, hidden from our sorrowing gaze, brought back in new birth, or perhaps in some empyrean continued, elevated above corruption, ruin, change, and disappointment? Such, at any rate, seems to be the almost universal instinctive belief of humans now and in the past. It seems a temptation written into our genes, this consoling or sometimes frightening belief, as is the very template of our capacity for language and personhood (Pinker, 1994). If so, that need not vouch for its truth, alas. Vitalism once seemed as true, as self-evident, as anything in the world or our experience of the world. It seemed as lasting and undeniable as . . . well, as the flatness of the earth, stretched to its four scriptural corners, circled by its tiny sun. Then we learned, with astonishment and resentment, that this flat cosmos is a small portion of a vast globe, that the small, brilliant, luminous ball crossing the sky is a sustained hydrogen-bomb explosion over a million kilometers broad, and that life is patterned matter. Vital force, élan vital, fell out of the scientific lexicon, along with the forces of the ancient astrologers who in their conceit took the vastly distant blazing stars for maps to local destiny. Can spiritlike powers have more durability than these lost guesses?

Paranormal phenomena seem to provide just the evidence we need to sustain spirit in the heart of matter, at least of human, conscious, loving, willing matter. Matter itself, after all, is not the hard, definitive stuff it was deemed to be a century ago before quantum theory. Matter and energy—the impulsive force that gives it motion and sustains its structure—have grown wispy and immaterial, a haze of quarks and stringy membranes, a dance of mathematical symmetries on an eleven-dimensional manifold no human eye is keen enough to witness nor human hand fine enough to shape. In an epoch of such drastic reinterpretation of spacetime and energy and matter, might we find an explanation, indeed a privileged place, for those rogue and damnable phenomena claimed anciently by witchcraft, formerly by spiritualism, and today by statistics-trained specialist researchers into canonically anomalous events and patterns? Might not psi be invited back within the gates of science?

We have found that trained remote viewers and ordinary folks alike can sometimes foresee with uncanny accuracy some unexpected future event, draw a convincingly detailed picture of some remote object hidden from sight, and deform the output of a random event generator. Perhaps not all of us can attain such gifts, even with expert guidance; after all, only a few can aspire to run a four-minute mile or calculate partial differential equations in our heads. Those are astonishing performances, yet hardly controversial, let alone paranormal. But suppose that sufficient candidates come forward for investigation by unprejudiced scientists, genuine psychics able to perform such feats repeatedly, perhaps astonishing even themselves. Some claim this as evidence of spirit, or at least of a breakdown in recognized scientific verities. Would it suffice to remake our scientific models and certainties?

Would it require those who respect evidence to readopt, whether grudgingly or gladly, a dualistic stance to the cosmos, to rediscover merits in a worldview that has been eroding for more than a century? Indeed, would it cause the traditionally religious to rethink their own views, confronting them with a miraculous order of reality that so many prelates have explained away as parables and allegories of a less educated time? Perhaps it seems obvious that people would hasten to embrace such support for long-held human yearnings. After all, some do so already, at the mere hint of miraculous manifestations, cures, or tears falling from the plaster or marble eyes of holy statues. The rest of us shake our sensible heads sadly. The psychologist James Alcock, long a skeptical critic of parapsychology, suggested in a special *Psi Wars* issue of the *Journal of Consciousness Studies* in 2003: "I continue to believe that parapsychology is, at bottom, motivated by belief in search of data, rather than data in search of explanation. It is the belief in a larger view of human personality and existence than is accorded to human beings by modern science that keeps parapsychology engaged in their search. Because of this belief, parapsychologists never really give the Null hypothesis a chance."

That kind of "will to believe," however, is not the postulated case under consideration. Suppose remarkable, inexplicable powers of mind were demonstrated, repeatedly and almost upon demand, as nowadays we take it for granted that the flick of a wall switch fills a dark room with fluorescent light brighter than any fire, and the turn of a key activates the quiet but immense power of an automobile engine stronger than any hero or beast of burden. Science and technology provided those technological benefits, the fruit of long, careful investigation of how things work, and of theories marked by increasing depth and extent, mapping the secret workings and patterns of the world. Would the acknowledged reality of psi in the midst of our workaday world oblige us to reshape our lives? Is a paranormal epiphany, even one we could turn on at will, the kind of experience we might expect to bring peace to the feuding, love to the loveless or hate-poisoned, food and comfort to the wretched, wisdom to the ignorant or puffed-up, and, most poignantly, authentic meaning to the well-off whose hearts and lives are empty?

It might seem so, for our legends, myths, and sacred teachings insist that teachers gifted with power and insight beyond the usual in a greedy, carnal world will prove their credentials by signs and wonders. Paradoxically, however, as we've noted, those same sources teach that miracles are unimportant, trivial, and distracting compared with heartfelt faith: an inward knowledge embraced precisely in the absence of empirical evidence or the scrutiny of logical reasoning. As Karl Popper, Imre Lakatos, and other philosophers of science teach, we must assess this claim with the greatest wariness. Although it can be found on the lips of the authentically wise, its shield also protects charlatans, the deluded, and the honestly mistaken who are wedded to their errors. More confrontingly still, we must acknowledge that the Western world's history is precisely unique in its characteristic methods of empirical science and materialistic technology. They summoned into routine reality a host of benefits (and some disasters) that a thousand years ago certainly would have been considered paranormal miracles. Meanwhile, most people in the Western world daily live more richly than kings of the ancient Orient. We might not be showered in gold and perfumes, but our deodorants are superior, our meals more various (if we take the trouble to avoid quick junk), and our knowledge far more extensive. Even without routine psi, we see the far places of the world at the touch of a switch, and our small machine deputies fling back to us images of other worlds from the edges of the solar system. This much is commonplace: we gaze upon atoms, or at the ancient universe nearly back to the big bang of creation. No other people has been as fortunate, as wealthy, as distracted by wonders. Perhaps distraction's lure is the key to what we are considering.

It cannot be astonishing wonders alone—whether technical or paranormal, whether inside or outside the gates of science—that will renew the world. Arguably the world already grows closer knit than ever before. If we are often distracted, alienated from loam, flower, beast, from collective passionate fellowship (corroboree, liturgy, even tribal warfare), the instruments of diversion serve also to show us the faces of those beyond our borders, the damage done by violence and natural disaster to our foes as well as to many people and creatures our ancestors never even knew existed. A nightclub or a skyscraper is torn by suicide bombers, and we see it; a village is bombed from the air, killing women and children; a town or county is lost to fire, flood, earthquake, or tsunami, and despite the fatigue of compassion, we do know some measure of empathy, and it grows a little more difficult for us to treat other people as irrelevant. Would access to genuine paranormal abilities augment the better face of information's Janus-blend: care for others versus hunger for base amusement?

Perhaps it would, or will. It is true that First Worlders can dial friends anywhere in the world on a mobile phone, our words shot nearly instantly via a satellite hung like a star in space, like whispering in a messenger angel's ear. Conceivably, true telepathy or vivid remote viewing might do what such technologies can never achieve: place us womb-deep within the consciousness of another. If it is true, as Roger Nelson hopes to establish, that random systems resonate to those agonies and triumphs shared by masses of humans—when we are brought to simultaneous focus, ironically, by those very instruments of technological communication—then perhaps we might finally cultivate a deeper fraternity. Yet we might still ask if these putative wonders of the mind must necessarily open an aperture into some superior, redemptive realm.

If telepathy, remote viewing, precognition, and psychokinesis become repeatably demonstrable, they will enter, at last, the realm of regular science, gifts granted us by rational inquiry, evidence, hard and imaginative thinking, explained by quantum theory or some new advanced physics. Will they—can they—yield in addition metaphysical insights, keys to happiness, intensity, existential meaning, a sea change in social direction? Only as much, perhaps, as a Cézanne rendering of a peach caught in light (which once brought me, motionless and timeless in front of a museum exhibit, to astonished tears), or a Bose-Einstein condensate of sodium gas poised a hair above absolute zero, or a Hubble portrait of the cosmic dawn of galaxies burning in illimitable night. Maybe paranormal phenomena will prove to be a gateway to some spiritual truth surpassing scientific knowledge. Even if it does not (which is my own estimate), the authenticity of psi—once proved—will stand as compelling, invigorating evidence that the world contains more than we have yet uncovered and explained. But this knowledge is already the true heart of science. Bit by bit, century by century, science replaces dogma with subtle understanding. Perhaps this will prove to be so in the case of those strange, intimate, persistent positive psi experiences most of us currently still feel obliged to name negatively, in our continuing partial ignorance, as *paranormal*.

ACKNOWLEDGMENTS

The effective exclusion of a challenging discipline like psi research beyond the barricaded gates of mainstream science has dismayed and annoyed me for decades. I hope this book helps ease open the gates at least a tad. It is not a textbook, nor is it meant to be read that way, but the inquiring reader can follow links to a fund of fascinating research that corroborates the story I tell. The generous help I've had from anomalies specialists (and more than one skeptical critic) while researching the book made the task possible. Despite the skeptical and uninformed opinion that parapsychology is a hollow quasi-science, a concoction of wishful thinking with no basis in experiment or theory, I'm here to tell you that researching the topic sometimes felt like being caught in a tornado of data, careful thinking, and disputatious opinion—some of it, admittedly, rather strange. I had the good fortune to be guided through this storm by patient experts, and I'm glad to acknowledge informative discussions with Eberhard Bauer, Dick Bierman, Stephen Braude, Richard Broughton, Jim Carpenter, York Dobyns, Suitbeit Ertel, Stanley Jeffers, Edwin May, Joe McMoneagle, Roger Nelson, Fotini Pallikari, Dean Radin, Mark Reilly, Ruth Reinsel, Stephan Schwartz, James Spottiswoode, Fiona

Steinkamp, Paul Stevens, Charley Tart, Robin Taylor, Russ Targ, Jessica Utts, the late Evan Harris Walker, and many others on several parapsychology lists. Hannah Jenkins at the University of Tasmania and Doug D'Elia in Southern California very generously provided me with important research materials, for which I am especially grateful. My warmest thanks go, as ever, to my dear wife, Barbara Lamar, for support, encouragement, and love.

Notes

Epigraphs

> "Decades of cumulating . . ." Dick J. Bierman, "On the Nature of Anomalous Phenomena . . ." 2001.

Chaper 1

Epigraph. Utts is citing Sybo Schouten, "Are we making progress?" in *Psi Research Methodology: A Re-examination, Proceedings of an International Conference, Oct 29–30, 1988,* edited by L. Coly and J. McMahon, New York: Parapsychology Foundation, Inc., 1993. Perhaps in the intervening twenty-odd years, the total has risen to as much as three months of research in conventional psychology.

1. Liza Gross, "Scientific Illiteracy and the Partisan Takeover of Biology," 2006.

2. Robert Jahn and Brenda Dunne, "The PEAR Proposition," 2005.

3. Helmut J. Schmidt, "PK Effect on Pre-Recorded Targets," 1976.

4. No significant results were found under either condition in Gertrude R. Schmeidler and Randall Borchardt, "Psi Scores with Random and Pseudo-Random Targets," in William G. Roll and John Beloff (eds.), *Research in Parapsychology 1980: Abstracts and Papers from the 23rd Annual Convention of the Parapsychological Association,* Metuchen, NJ: Scarecrow Press, 1981. For PEAR, see R. G. Jahn,B. J. Dunne, and R. D. Nelson, "Engineering Anomalies Research," *Journal of Scientific Exploration,* vol. 1, no. 1 (1987), 21–50: "In particular, it is reasonable to ask whether the physical behavior of the noise source itself is affected during the PK efforts, and if so, in what way.... The results of 29 experimental series employing this pseudo-random source are also statistically significant with a probability

against chance of .003 . . . and the individual operator signatures show strong qualitative similarities to those achieved on the standard REG." On the other hand, York Dobyns of PEAR told me: "What we had seen, and published as of a few years ago, is that pseudo-random sources appear to have no effects" (personal communication, October 2, 2006). What of Jahn's earlier claim? It turns out that it was a design error, uncovered in the late 1980s. "Although the source described there was intended to be pseudo-random, it was found later that it could not be set to a consistent starting state in which the results would be the same. The culprit was a variable-rate sampling system which introduced physical noise via the clock jitter." That is, the REG was actually spewing out true noise-driven random numbers after all.

5. Stanley Jeffers, "The PEAR Proposition—Fact or Fallacy?" 2006.

6. York Dobyns, "Overview of Several Theoretical Models on PEAR Data," 2000.

Chapter 2

1. Po Bronson, "A Prayer before Dying," http://www.pobronson.com/A_Prayer_Before_Dying.htm.

2. Ingo Swann, "Remote Viewing, the Real Story: An Autobiographical Memoir," chapter 28, "My First Letter to Dr. H. E. Puthoff, March 1972," Superpowers of the Human Biomind, 1996, http://www.biomindsuperpowers.com/Pages/RealStoryCh28.html.

3. This stage-managed and sorry business is discussed in some detail in my book about parapsychology, *The Lotto Effect: Towards a Technology of the Paranormal*, Hawthorn, AU: Hudson, 1992.

Chapter 3

1. Stephan A. Schwartz Web site, "Papers and Research Reports," http://www.stephanaschwartz.com/home.htm.

2. Edgar Cayce, "Earth Changes," http://www.crystalinks.com/caycearthchanges.html.

3. "Susan Smith: Child Murderer or Victim?" "Court TV Crime Library, http://www.crimelibrary.com/notorious_murders/famous/smith/.

4. These documents can be downloaded from http://www.mod.uk/DefenceInternet/FreedomOfInformation/DisclosureLog/SearchDisclosureLog/RemoteViewing.htm, which is the source of the summary information that follows.

Chapter 4

1. Edwin C. May, et al., "Anomalous Anticipatory Skin Conductance Response to Acoustic Stimuli: Experimental Results and Speculation About a Mechanism," *Journal of Alternative and Complementary Medicine*, vol. 11, no. 4 (2005): 695–702.

2. Edwin C. May, S. James P. Spottiswoode, and Christine L. James, "Shannon Entropy: A Possible Intrinsic Target Property," 1994.

3. Randi's misleading criticisms of some psi studies are discussed in some detail in my *The Lotto Effect*.

4. Dick J. Bierman, "On the Nature of Anomalous Phenomena," 2001. It is necessary to add that a PK meta-analysis by Bösch, Steinkamp, and Boller ("Examining Psychokinesis," 2006) shows only a very small effect size, and the authors conclude: "The meta-analysis combined 380 studies that assessed whether RNG output correlated with human intention and found a significant but very small overall effect size. The study effect sizes were strongly and inversely related to sample size and were extremely heterogeneous. A Monte Carlo simulation revealed that the small effect size, the relation between sample size and effect size, and the extreme effect size heterogeneity found could in principle be a result of publication bias." Authors of earlier, positive meta-analyses vehemently disagreed; see Radin, Nelson, Dobyns, and Houtkooper, "Reexamining Psychokinesis" (2006).

Chapter 5

1. Mark Henderson, "Theories of Telepathy and Afterlife Cause Uproar at Top Science Forum," September 6, 2006, http://www.timesonline.co.uk/article/0,,2-2344804,00.html.

2. Rupert Sheldrake, "Gosh, I Was Just Thinking about You," Comment section, September 7, 2006, http://www.timesonline.co.uk/article/0,,6-2346084,00.html.

3. "BA Festival of Science Discussion between Rupert and Prof. Peter Atkins," BBC Radio Five Live, September 6, 2006: http://www.sheldrake.org/D&C/controversies/Atkins_discussion.html.

4. Brian Josephson, "Physics and the Nobel Prizes, (article included in a booklet accompanying the Royal Mail special stamps issued on October 2, 2001 to commemorate the centenary of the Nobel prizes)," http://www.tcm.phy.cam.ac.uk/~bdj10/stamps/text.html.

5. Erica Larriech, "Stamp Booklet Has Physicists Licked," *Nature* 413, no. 339 (2001), http://www.tcm.phy.cam.ac.uk/~bdj10/stamps/nature.html.

6. Brian Josephson, "Telepathy Has Stamp of Truth: The Big Issue," *The Observer,* Letters section, October 7, 2001, http://observer.guardian.co.uk/letters/story/0, 6903,564641,00.html. Josephson and Pallikari's paper is actually from 1991: http://www.tcm.phy.cam.ac.uk/~bdj10/papers/bell.html. See also Professor Josephson's "Beyond Quantum Theory" (2002): "Physics, in advocating quantum mechanics as a basis for a 'theory of everything,' may have moved too fast towards a too tempting conclusion, and thrown out the crucial and subtle intelligence of the observer as a part of this process." Josephson's "String

Theory, Universal Mind, and the Paranormal" (2003) attempts to explain ESP "in terms of shared 'thought bubbles' generated by the participants out of the mental vacuum state." Despite the paper's title, it concludes with the admission that "Since our proposals (such as thought bubbles emerging from some kind of background) do not involve the precise details of string theory, they may survive any such changes that fundamental science may undergo."

7.   "The Telepathy Debate, Royal Society of Arts, London, 15th January 2004." Available at Skeptical Investigations, http://www.skepticalinvestigations.org/whoswho/telepathy_RSA.htm.

8.   Robert G. Jahn and Brenda J. Dunne, "Sensors, Filters, and the Source of Reality," *Journal of Scientific Exploration* 18, no. 4 (2004): 547–70. Available at http://www.astrosciences.info/18.4_jahn_dunne.pdf.

9.   Topics such as quantum mechanics, quantum theory, and the far more demanding quantum field theory are all addressed in Wikipedia. See http://en.wikipedia.org/wiki/Quantum_mechanics, http://en.wikipedia.org/wiki/Quantum_theory, and the far more demanding http://en.wikipedia.org/wiki/Quantum_field_theory.

10.   Alex Vilenkin, "The Principle of Mediocrity," excerpt from *Many Worlds Are One* (Hill and Wang, 2006), http://www.edge.org/3rd_culture/vilenkin06/vilenkin06_index.html.

11.   http://www.wddty.co.uk/thefield/noflash/index.asp.

12.   Stephen G. Brush, "The Chimerical Cat: Philosophy of Quantum Mechanics in Historical Perspective," *Social Studies of Science* 10, no. 4, 393–447.

13.   Philip Ball, "Hawking Rewrites History . . . Backwards," 2006.

14.   Scott LaFee, "Cause and Defect," *San Diego Union-Tribune,* June 22, 2006. Available at http://www.signonsandiego.com/news/science/20060622-9999-lz1c22cause.html.

Chapter 6

1.   Personal Communication, September 20, 2006. J. W. Hartwell, "A Bound for the Observational Theories of Psi," 1977. Also B. Millar, and J. Hartwell, "Dealing with Divergence," 1979.

2.   Paul C. W. Davies, "Quantum Fluctuations and Life," abstract, http://arxiv.org/pdf/quant-ph/0403017.

Chapter 7

Epigraph. Edwin C. May, Jessica M. Utts, and S. James P. Spottiswoode, "Decision Augmentation Theory: Toward a Model of Anomalous Mental Phenomena," 1995.

1.   Dick J. Bierman, "The PRL Autoganzfeld Revisited: Refuting the Sound Leakage

Hypothesis," *Journal of Parapsychology* 63 (Sept. 1999): 271–74. Available at http://m0134.fmg.uva.nl/publications/1999/soundleakage_JoPSept99.pdf.

2. Greg Egan, *Teranesia: A Novel* (New York: HarperPrism, 1999), see http://gregegan.customer.netspace.net.au/TERANESIA/TERANESIA.html.

3. See Johnjoe McFadden's Web site at http://www.surrey.ac.uk/qe/.

4. Damien Broderick, "The Ballad of Bowsprit Bear's Stead," Excerpt and e-book available at Fictionwise, http://www.fictionwise.com/ebooks/eBook539.htm.

5. Robin Taylor, "Evolutionary Theory and Psi," 2003.

6. Edgar Mitchell, Introduction to Dale E. Graff, *Tracks in the Psychic Wilderness*, 1998.

7. Paola Palladino and Rossana De Beni, "When Mental Images Are Very Detailed," 2003.

Chapter 8

1. H. Häffner, F. Schmidt-Kaler, et al., "Robust Entanglement," http://arxiv.org/pdf/quant-ph/0508021.

2. Report of work by Vlatko Vedral et al., *Phys. Rev. Lett.* 96 060407, in "Entanglement Heats Up," PhysicsWeb, 23 February 2006, http://physicsweb.org/articles/news/10/2/14/1.

3. Edwin C. May, Jessica M. Utts, and S. James P. Spottiswoode, "Decision Augmentation Theory," 1995.

4. "Swedenborgianism," Christian Apologetics and Research Ministry, http://www.carm.org/list/swedenborg.htm.

5. For example, York H. Dobyns and R. D. Nelson, "Empirical Evidence against Decision Augmentation Theory," 1998, and York H. Dobyns, "Overview of Several Theoretical Models on PEAR Data," 2000.

Chapter 9

1. I have published two linked novels, *Godplayers* (New York: Thunder's Mouth, 2005) and *K-Machines* (New York: Thunder's Mouth, 2006), that playfully explore the consequences of life in such a computational cosmos for those with a privileged access to its program.

2. Charles Tart, "Life in the World Simulator," 1990. See also "Multiple Personality, Altered States and Virtual Reality," 1991.

3. Frederik Pohl, "The Tunnel Under the World," *Galaxy* magazine, January 1955.

4. See the Wikipedia entry for Simulacron-3 at http://en.wikipedia.org/wiki/Simulacron-3.

5. See, for example, Gerald A. Larue, professor emeritus of religion and adjunct professor of gerontology, University of Southern California, "Afterlife," 1989.

GLOSSARY

AGENT: In classic parapsychological models, the person whose states of feeling or consciousness are to be apprehended *paranormally* by the psi *percipient*.

ANOMALOUS COGNITION (AC): *ESP*

ANOMALOUS PERTURBATION (AP): *PK*

BIOPHOTONS: Alleged ultraweak light emitted from and within living systems.

CALL: The subject's recorded cognitive response to a target not knowable via ordinary means.

CHANCE: The sum of normal causative but untrackable factors (other than those produced by intention) that influence a situation.

CLAIRVOYANCE: Psi-mediated perception at a distance.

CONFABULATION: Active process of confusing memory, intention and imagination; also, of unconsciously contriving imaginary explanations and motives for puzzling experiences or choices.

CSICOP/CSI: Premier skeptical organization, formerly the Committee for the Scientific Investigation of Claims of the Paranormal, now Committee for Skeptical Inquiry.

DECISION AUGMENTATION THEORY (DAT): A theory of psi in which humans integrate psychic information into ordinary decisions; for example, by using precognition to select an optimal starting point in an otherwise random sequence.

DECOHERENCE: The condition whereby an isolated superposed quantum system becomes connected to its larger context and loses its quantum weirdness.

DELUSIONAL REINFORCEMENT: The process of reinforcing a delusion through the encouragement of others.

DEVIATION: The extent to which the observed score differs from the *Mean Chance Expectation*, in either direction—that is, high or low.

DMILS: Direct mental interaction with living systems.

DIFFERENTIAL EFFECT: Statistically significant differences in scoring direction when experiments or psi applications are presented in two contrasted conditions; for example, when pictorial targets are interspersed with verbal targets. It has been claimed repeatedly that such deliberate contrasts can "force" respondents—with a certain degree of reliability—to score above average on one class of targets and below average on the other. A second form of differential effect has been found when respondents are sorted by gender, mood or some psychological parameter such as introversion and extraversion, or credulity versus skepticism. Careful attention to such effects can perhaps turn *psi mode* variation to advantage, segregating *psi-hitting* into one group of calls and *psi-missing* into the remainder.

EGGs (electroGaiagraphs): The whimsical name given to random number generators suspected of being influenced by human states of consciousness on a global scale.

ESP (EXTRASENSORY PERCEPTION): An affective or cognitive response (that is, either of feeling or thought/image) to some target or circumstance not knowable by normal means.

ERV (extended remote viewing): One of the many brand-name variants of remote viewing protocols developed by the formerly classified US government project known as Star Gate.

ENSP (Evolution's Need Serving Psi): A model of psi devised by Dr. Robin Taylor with emphasis on its biological relevance and presumed evolutionary history.

EXAPTATION: A kind of evolutionary change in which a biological system selected for one purpose has been repurposed for another.

EXPERIMENTER EFFECT: One intriguing correlation to emerge from parapsychological experiments is the interaction between experimenter and putative subject. Recently, it has been noticed that particular experimenters seem to elicit either strong or null results from subjects, even when all other controls are retained in identical condition. This has given rise to the notion that psychic phenomena are always significantly interactive, that there is no possibility of a purely objective stance for an experimenter, as there might be in a physics experiment.

EXTINCTION PARADIGM: An experimental protocol that accidentally rewards failure rather than success.

fMRI (FUNCTIONAL MAGNETIC RESONANCE IMAGING): A form of laboratory/medical scanner that non-intrusively detects the relationship between changes in blood flow, for example, and mental activity of quite specific kinds.

GANZFELD PSI (GF or sometimes GZ): A technique intended to enhance the psi function by reducing sensory input to a bland minimum, without going all the way to sensory deprivation (which actually increases "internal noise"). It has been one of the most successful means of building a credible database of ESP events, and the site of a fruitful debate between parapsychology and its critics.

GLOBAL CONSCIOUSNESS PROJECT (GCP): A project initiated by scientists at *PEAR*, intended to track and notate any systematic changes in a network of random number generators scattered around the world and not connected electronically.

INDEX RUN: A small set of randomly interleaved targets, decoded in advance of the bulk of a psi application's results, that can detect—and thus help compensate for—*psi mode* and idiosyncratic response biases manifested in that particular application.

LST (Local Sidereal Time): Local time measured by the apparent motion of the stars rather than by the apparent daily motion of the Sun.

MAJORITY-VOTE TECHNIQUE: A method for concentrating the fleeting and unpredictable impact of psi events by making many attempts at a paranormal performance, accepting as the response that option chosen more often than any other. Because of response bias and various preference factors, it is often necessary to *normalize* the accumulated scores before seeking the majority vote. For example, if the task is to guess a number in the range 1 through 10, 100 guesses might be made. In theory, each candidate number might be expected to score close to 10 apiece. Because of widespread preferences, however, it is very likely that 7 will be chosen far more often than that (and 1 or 10 far less). Suppose the typical vote for 7 is always close to 30 (rather than the 10 we'd get from a randomizing machine). A simple majority vote would therefore nearly always select 7 as the paranormal response. To overcome such biases, it is essential to establish a baseline of preferences, and then—for example—divide each score by its typical weight. These normalized data can thereafter be scrutinized for the majority vote. Results from remote viewing experiments suggest that deliberate redundancy of this kind can lead to concentrated error just as easily as to concentrated accuracy (perhaps due to a kind of psychic leakage between the tasked viewers).

MEAN CHANCE EXPECTATION (MCE): Average score in a controlled test when only chance is operative.

MEDIUMS (CHANNELS): Traditional and more recent terms for people who believe they can contact the dead and bring information to the living.

MODEL OF PRAGMATIC INFORMATION (MPI): A model of psi devised by Dr. Walter von Lucadou, emphasizing the meaning of paranormal information to a psychic system.

NORMALIZATION: Any method that brings sets of data into overall conformity with each other so they might be fairly compared.

OUT OF BODY EXPERIENCE (OOBE, OBE, OB): Vivid hallucinations of one's "point of consciousness" being separated from the body, though perhaps connected by a cord, and able to move "with the speed of thought" to various locations in the real world and even beyond it. Cognitive science offers a persuasive explanation for this experience in terms of brief disruptions of the mental models we use to position ourselves in space and time, although it has been claimed that some validated paranormal information has been obtained in this state.

PARANORMAL: Hypothetically, intentional effects beyond the ability of current scientific paradigms (or disciplinary frameworks) to explain or replicate.

PARAPSYCHOLOGY: The science (or pre-science) using the methods of accepted research disciplines in the established sciences to investigate claims of paranormal perception (*ESP*) and action (*PK*).

PEAR: Princeton University's long-running Engineering Anomalies Research program, now closed down, headed by Professor Robert Jahn and psychologist Brenda Dunne.

PERCIPIENT: The subject of cognitive or affective paranormal phenomena.

PK (PSYCHOKINESIS): The unmediated paranormal impact of intention on the external world. This term, introduced by J. B. Rhine to replace "telekinesis" which was tainted by spiritualism, contains an unfortunate theoretical loading of its own. Alternatives are "conformal behavior" and "anomalous perturbation."

PMIR (Psi-Mediated Instrumental Response): A variant conceptualization of PK, devised by Dr. Rex Stanford.

POLTERGEIST: Traditionally, a mischievous sprite responsible for the anomalous physical effects of a supposed haunting. See *RSPK*.

PRECOGNITION: Statistically significant non-inferential knowledge-claims of future random or otherwise unpredictable situations.

PRECOGNITIVE REMOTE PERCEPTION: Remote perception of events that have not yet taken place.

PRESENTIMENT: See PRESTIMULUS RESPONSE.

PRESTIMULUS RESPONSE (PSR): A measurable change of response to a shocking, arousing, or neutral stimulus in advance of the stimulus being selected and applied.

PRIMES: Orienting experiences that make it easier to choose one path rather than another.

PROBABILITY FIELD: A conjectural "field" of probability akin to an electric, magnetic, or gravitational field.

PSI: A general term embracing putative paranormal links between a person and the external world (including other people).

PSI-HITTING: Hypothetical explanation for a succession of trials in which results are statistically more accurate than *mean chance expectation* (preferably at such a level that the chance of coincidence accounting for the excess successes are 1 in 100 or less).

PSI-MISSING: The reverse of psi-hitting—except that a significant deficit of correct calls is found. This category runs distinctly counter to commonsense, and perhaps is best grasped by analogy to a slip of the tongue: there, the particular wrong word is uttered only because the right word is "on the tip of your tongue," although some unconscious error or motive switches wrong for right. To allow for the presence of psi-missing (found in many parapsychological experimental results), most psi studies evaluate their results according to a two-tailed estimate of *significance*—a more stringent measure than usual, explained in all elementary statistics texts.

PSI MODE: A hypothetical condition within a given experiment or psi application favoring *psi-hitting* or *psi-missing*.

PSYCHIC: Attributable to *psi*.

QUANTUM ZENO EFFECT: The weird quantum effect whereby an unstable particle, if observed continuously, will never decay.

QUBIT: A unit of superposed quantum information that is both one and zero simultaneously.

RANDOM EVENT GENERATOR (REG) or RANDOM NUMBER GENERATOR (RNG): An electronic device producing uncorrelated sequences of ones and zeros.

RECURRENT SPONTANEOUS PSYCHOKINESIS (RSPK): An alternative term for the traditional *poltergeist,* that assumes the cause of anomalous activity is a living human person attached to the situation.

REDUNDANCY: Repetition of a task.

REMOTE VIEWING: A formal process of applied *clairvoyance,* preferably with double blinded evaluation by judges otherwise unconnected with the target options.

SCIENCE APPLICATIONS INTERNATIONAL CORPORATION (SAIC): The largest employee-owned research and engineering firm in the United States, heavily supported by the United States Department of Defense and the Intelligence Community, including the National Security Agency.

SCIENTOLOGY: A cult started by science fiction writer L. Ron Hubbard, somewhat influential in early development of remote viewing.

SGRA*(SAGITTARIUS A STAR): A massively energetic source near the centre of the Milky Way galaxy.

SHANNON ENTROPY: A formal mathematical measure of information.

SIDEREAL TIME: see *LST*.

SIGNIFICANCE: A numerical result attains statistical significance when it equals or exceeds a given criterion of chance improbability. The *Journal of Parapsychology* (from whose glossary some of these definitions have been adapted), declares: "The criterion commonly used in parapsychology today is a probability value of .02 (odds of 50 to 1 against chance) or less, or a deviation in either direction such that the [number of *standard deviations*] is 2.33 or greater. Odds of 20 to 1 (probability of .05) are regarded as strongly suggestive." Even this authoritative source gets it slightly wrong: the probability of .02 is actually equal to odds of 49 to 1, while .05 equals odds of 19 to 1. (In the first case, there are 49 ways in 50 of being wrong, and only 1 way of being right.) What's more, because the paranormal hypothesis is so extremely hard to swallow, these days the preferred significance level is more likely to be .01—a scant 1 chance in 100 of simple coincidence—or even less.

SIGNATURES: Recurrent patterns discerned by some in the results generated by at least some agents in psychic tasks, such as those favoured by PEAR.

STACKING EFFECT: Spurious high or low results can easily be found in psi applications using repeated guesses or efforts, either by one person or a group, at a restricted list of targets. In essence, such protocols run foul of mass preferences and response biases for and against certain options.

STANDARD DEVIATION: Usually the "root mean square" of all the individual deviations. A handy formula is "find the square root of (the Number of Trials [or guesses], multiplied by the probability of guessing right, multiplied by the probability of guessing wrong)".

STANFORD RESEARCH INSTITUTE, INTERNATIONAL (SRI or SRII): A Californian think tank situated in Menlo Park, CA.

STAR GATE: The most recent code name of the formerly secret psi research program conducted by the United States government.

STEREOPSIS: The capacity to see depth by combining two slightly separated images, one from each eye.

SUBJECTIVE VALIDATION: Acceptance of paranormal claims on the basis of descriptions vague and general enough to apply to a wide range of people and situations.

SUPERPOSITION: In the quantum realm of the very small, particles remain in a condition that comprises all possible states of the particles.

TARGET: In ESP tests, a shielded or as-yet-non-existent situation to which a percipient is trying to respond. In PK tests, an equivalent arrangement the respondent is trying to influence.

TELEPATHY: Paranormal perception of the thoughts or feelings of another person.

THETAN: The term devised by Scientology mythology for the supposedly immortal part of a person, brought to earth tens of millions of years ago by a galactic dictator.

TOOTH FAIRY: Cute creature who swaps children's lost teeth (carefully positioned under the pillow) for coins, or, these days, folding money or Gold American Express cards. Frequently misused by snide skeptics as an analogy with parapsychology's claims for psi. Unlike psi phenomena, however, Tooth Fairy (and the cognate Santa Claus phenomena) are generally agreed to have been satisfactorily explained in orthodox terms.

WAVE FUNCTION: The mathematical tool that quantum mechanics uses to describe any physical system. It is a function from a space that consists of the possible states of the system.

WEAK QUANTUM THEORY: A model of quantum theory that employs an algebraic axiomatic approach using only the most basic and general formal structures characteristic of quantum mechanics, while omitting more specific definitions.

ZERO-POINT FIELD THEORY (ZPF): A theory that a background sea of quantum light fills the universe, generating a force (inertia) that opposes acceleration when any material object is pushed.

BIBLIOGRAPHY

This list of references is not meant to be exhaustive. A complete bibliography of publications dealing seriously with anomalous phenomena would be many thousands of pages long.

Alcock, James, Jean Burns, and Anthony Freeman, eds. *Psi Wars: Getting to Grips with the Paranormal.* Special Double Issue of *Journal of Consciousness Studies* 10, no. 6–7 (June–July 2003).

Atmanspacher, H., H. Romer, and H. Walach. "Weak Quantum Theory: Complementarity and Entanglement in Physics and Beyond." *Foundations of Physics* 32, no.3 (2002): 379–406.

Ball, Philip. "Hawking Rewrites History . . . Backwards: To Understand the Universe We Must Start from the Here and Now." *Nature* (June 21, 2006). Available at http://news.nature.com//news/2006/06 0619/060619-6.html.

Barbour, Julian. *The End of Time: The Next Revolution in Our Understanding of the Universe.* London: Weidenfeld and Nicolson, 1999.

Barrow, John. *Impossibility: The Limits of Science and the Science of Limits.* New York: Oxford University Press, 1998.

Bem, D. J., and C. Honorton. "Does Psi Exist? Replicable Evidence for an Anomalous Process of Information Transfer." *Psychological Bulletin* 115, no. 1 (1994): 4–18.

Bem, D. J., J. Palmer, and R. S. Broughton. "Updating the Ganzfeld Database: A Victim of Its Own Success?" *Journal of Parapsychology* 65(2001): 207–18.

Bieler, Peter, with Suzanne Costas. *"this business has legs": How I Used Informercial Marketing to Create the $100,000,000 THIGHMASTER Craze: An Entrepreneurial Adventure Story.* New York: John Wiley and Sons, 1996.

Bierman, Dick J. "Anomalous Baseline Effects in Mainstream Emotion Research Using Psychophysiological Variables." *Proceedings of Presented Papers: The 43rd Annual Convention of the Parapsychological Association* (2000): 34–47.

———. "On the Nature of Anomalous Phenomena: Another Reality between the World of Subjective Consciousness and the Objective World of Physics?" In *The Physical Nature of Consciousness,* ed. Philip van Loocke. Philadelphia, PA: John Benjamins, 2001.

Blackmore, Susan. "The Lure of the Paranormal." *New Scientist* 127, no. 1735 (September 22, 1990): 62–65.

Boller, Emil, Holger Bösch, and Fiona Steinkamp. "Experiments Examining the Possibility of Human Intention Interacting with Random Number Generators: A Preliminary Meta-Analysis." *Journal of Parapsychology* 66 (September 1, 2002).

Bösch, Holger, Fiona Steinkamp, and Emil Boller. "Examining Psychokinesis: The Interaction of Human Intention with Random Number Generators: A Meta-Analysis." *Psychological Bulletin* 132, no. 4 (2006): 497–523.

Brian, Denis. *The Enchanted Voyager: The Life of J. B. Rhine.* Englewood Cliffs, NJ: Prentice-Hall, 1982.

Broderick, Damien. *The Lotto Effect: Towards a Technology of the Paranormal,* Hawthorn, AU: Hudson,1992.

Broughton, Richard S. "If you want to know how it works, first find out what it's for." In Debra H. Weiner and Robert L. Morris, eds. *Research in Parapsychology 1987.* The Scarecrow Press, Metuchen, N.J. and London, 1988: 187–202.

Cardeña, Etzel, Stephen J. Lynn, and Stanley Krippner. *Varieties of Anomalous Experience: Examining the Scientific Evidence.* Washington, DC: American Psychological Association, 2000.

Carter, John. *Sex and Rockets: The Occult World of Jack Parsons.* Venice, CA: Feral House, 1999.

Cramer, John G. "An Overview of the Transactional Interpretation of Quantum Mechanics." *International Journal of Theoretical Physics* 27, no. 2 (1988): 227–36.

Dobyns, York H. "Selection versus Influence in Remote REG Anomalies." *Journal of Scientific Exploration* 7, no. 3 (1993): 259–69.

————. "Overview of Several Theoretical Models on PEAR Data." *Journal of Scientific Exploration* 14, no. 2 (2000): 163–94. Available at http://www.princeton.edu/~pear/pdfs/jse_papers/Overview.pdf.

Dobyns, York H., and R. D. Nelson. "Empirical Evidence against Decision Augmentation Theory." *Journal of Scientific Exploration* 12, no. 2 (1998): 231–58. Available at http://www.princeton.edu/~pear/pdfs/jse_papers/Evidence-against-DAT.pdf.

Druckman, Daniel, and John A. Swets, eds. *Enhancing Human Performance: Issues, Theories, and Techniques.* Washington, DC: National Academy Press, 1988.

DuBois, Allison. *Don't Kiss Them Good-bye.* New York: Fireside/ Simon and Schuster, 2005.

Dunne, B. J., and R. G. Jahn. "Information and Uncertainty in Remote Perception Research." *Journal of Scientific Exploration* 17, no. 2 (2003): 207–41.

Dunne, B. J., R. D. Nelson, Y. H. Dobyns, and R. G. Jahn. "Individual Operator Contributions in Large Data Base Anomalies Experiments." Technical Note PEAR 88002. Princeton University School of Engineering and Applied Science, July 1988.

Eysenck, Hans J., and Carl Sargent. *Explaining the Unexplained: Mysteries of the Paranormal.* London: Weidenfeld and Nicolson, 1982.

Feinberg, Gerald. "Possibility of Faster-Than-Light Particles." *Physical Review* 159, no. 5 (1967): 1089–1105.

———. "Precognition: A Memory of Things Future." In Oteri, *Quantum Physics and Parapsychology,* 54–64.

Flew, Antony. *Merely Mortal? Can You Survive Your Own Death?* Amherst, NY: Prometheus Books, 2000.

Frayn, Michael. *The Human Touch: Our Part in the Creation of a Universe.* London: Faber, 2006.

Frazier, Kendrick, ed. *Science Confronts the Paranormal.* Buffalo, NY: Prometheus Books, 1986.

Gardner, Martin. *Weird Water and Fuzzy Logic: More Notes of a Fringe Watcher.* Amherst, NY: Prometheus, 1996.

Graff, Dale E. *Tracks in the Psychic Wilderness: An Exploration of ESP, Remote Viewing, Precognitive Dreaming, and Synchronicity.* Rockport, MA: Element Books, 1998.

Gross, Liza. "Scientific Illiteracy and the Partisan Takeover of Biology." *Public Library of Science* Biology 4, no. 5 (April 18, 2006): 680–83. Available at http://biology.plosjournals.org/perlserv?request =get-document&doi=10.1371/journal.pbio.0040167.

Hansel, C. E. M. *The Search for Psychic Power: ESP and Parapsychology Revisited.* Buffalo, NY: Prometheus, 1989.

Hansen, George. *The Trickster and the Paranormal.* Philadelphia: Xlibris, 2001.

Harary, Keith. "Selling the Mind Short: The Myth of Psychic Privilege." *Omni,* April 1994: 6. Unauthorized copy available at http://zayexi.tripod.com/article.html.

———. "Confessions of a Star Psychic." *Psychology Today,* November/December 2005, 76–84.

Hartwell, J. W. "A Bound for the Observational Theories of Psi." *European Journal of Parapsychology* 2, no. 1 (1977): 19–28.

Honorton, C., and D. C. Ferrari. "Future Telling: A Meta-Analysis of Forced-Choice Precognition Experiments, 1935–1987." *Journal of Parapsychology* 53 (1989): 281–308.

Horrigan, Bonnie J. "Profile of Stephan A Schwartz: The Realm of the Will." *Explore,* May 2005. Available at http://www.stephanaschwartz.com/PDF/StephanSchwartz.pdf.

Houtkooper, J. M. "Arguing for an Observational Theory of Paranormal Phenomena." *Journal of Scientific Exploration* 16, no. 2 (2002): 171–85.

Hyman, Ray. "The Ganzfeld Psi Experiment: A Critical Appraisal." *Journal of Parapsychology* 49, no. 11 (1985): 3–49.

———. "Parapsychological Research: A Tutorial Review and Critical Appraisal." Invited Paper. *Proceedings of the IEEE* 74, no. 6 (1986): 823–49.

———. "Evaluation of Program on Anomalous Mental Phenomena." 1995. Available at http://www.mceagle.com/remote-viewing/refs/science/air/hyman.html.

Hyman, Ray, and Charles Honorton. "A Joint Communique: The Psi Ganzfeld Controversy." In Ganzfeld Debate Responses, Special Number of *Journal of Parapsychology* 50, no. 4 (December 1986): 351–64.

Inglis, Brian. *Natural and Supernatural: A History of the Paranormal from Earliest Times to 1914.* London: Hodder and Stoughton, 1977.

————. *Science and Parascience: A History of the Paranormal, 1914–1939.* London: Hodder and Stoughton, 1984.

Jahn, Robert G. "Anomalies: Analysis and Aesthetics," *Journal of Scientific Exploration* 3, no. 1 (1989): 15–26.

Jahn, Robert G., and Brenda J. Dunne. *Margins of Reality: The Role of Consciousness in the Physical World.* San Diego: Harcourt Brace Jovanovich, 1987.

————. "The PEAR Proposition." *Journal of Scientific Exploration* 19, no. 2 (2005): 195–245.

Jeffers, S. "The PEAR Proposition: Fact or Fallacy?" *Skeptical Inquirer* 30, no. 3 (May/June 2006).

Josephson, Brian. "Beyond Quantum Theory: 'A Realist Psycho-Biological Interpretation of Reality' Revisited." *Biosystems* 64, no. 1–3 (January 2002): 43–45. Available at http://arxiv.org/ftp/quant-ph/papers/0105/0105027.pdf.

————. "String Theory, Universal Mind, and the Paranormal." 2003. Available at http://arxiv.org/html/physics/0312012.

Keil, Jürgen, ed. *Gaither Pratt: A Life for Parapsychology.* Jefferson, NC: McFarland, 1987.

Kress, Kenneth A. "Parapsychology in Intelligence: A Personal Review and Conclusions." *Journal of Scientific Exploration* 13, no. 1 (1999): 69–87. Originally published in the Winter 1977 issue of *Studies in*

Intelligence, a classified CIA internal publication; released to the public in 1996. Available at: http://www.scientificexploration. org/jse/articles/pdf/13.1_kress.pdf.

Krippner, Stanley. *Song of the Siren: A Parapsychological Odyssey.* New York: Harper and Row, 1975.

Krippner, Stanley, ed. *Advances in Parapsychological Research,* vol. 5. Jefferson, NC: McFarland, 1987.

———. *Advances in Parapsychological Research,* vol. 6. Jefferson, NC: McFarland, 1990.

Kurtz, Paul, ed. *Skeptical Odysseys: Personal Accounts by the World's Leading Paranormal Inquirers.* Amherst, NY: Prometheus Books, 2001.

Larue, Gerald A. "Afterlife." *Humanism Today* 5 (1989): 36–61. Available at http://www.humanismtoday.org/vol5/larue.pdf.

Lloyd, Seth. *Programming the Universe: A Quantum Computer Scientist Takes on the Cosmos.* New York: Knopf, 2006.

Mandelbaum, W. Adam. *The Psychic Battlefield: A History of the Military-Occult Complex.* New York: St Martin's Press, 2000.

Marks, David F. "Remote Viewing Revisited," *Skeptical Inquirer* 6 (4) 18–29, 1986.

Marks, David, and Richard Kammann. *The Psychology of the Psychic.* Buffalo, NY: Prometheus Books, 1980.

May, Edwin C., S. James P. Spottiswoode, and Christine L. James. "Shannon Entropy: A Possible Intrinsic Target Property." *Journal of Parapsychology* 58, no. 4 (December 1994): 384–401.

May, Edwin C., Jessica M. Utts, and S. James P. Spottiswoode. "Decision Augmentation Theory: Toward a Model of Anomalous Mental

Phenomena." *Journal of Parapsychology* 59 (September 1995). Available at http://www.jsasoc.com/docs/DAT-I.pdf.

McCaffrey, Anne. *The Wings of Pegasus.* New York: Guild America Books, 1973. (Includes "A Womanly Talent," previously published in *Analog* 82, no. 6 [February 1969].)

McClenon, James. *Deviant Science: The Case of Parapsychology.* Philadelphia: University of Pennsylvania Press, 1984.

McKie, Robin. "Royal Mail's Nobel guru in telepathy row." *The Observer*, September 30, 2001. Available at http://observer.guardian.co.uk/uk_news/story/0,6903,560604,00.html.

McMahon, David. *Quantum Mechanics Demystified: A Self-Teaching Guide.* New York: McGraw-Hill, 2006.

McMoneagle, Joseph. *The Ultimate Time Machine: A Remote Viewer's Perception of Time and Predictions for the New Millennium.* Charlottesville, VA: Hampton Roads, 1998.

————. *Remote Viewing Secrets: A Handbook.* Charlottesville, VA: Hampton Roads, 2000.

————. *The Stargate Chronicles: Memoirs of a Psychic Spy.* Charlottesville, VA: Hampton Roads, 2002.

McTaggart, Lynne. *The Field: The Quest for the Secret Force of the Universe.* New York: HarperCollins, 2002.

Millar, B., and J. Hartwell. "Dealing with Divergence." In *Research in Parapsychology 1978,* ed. W. C. Roll, 91–93. Metuchen, NJ: Scarecrow Press, 1979.

Milton, J., and R. Wiseman. "Does Psi Exist? Lack of Replication of an Anomalous Process of Information Transfer." *Psychological Bulletin* 125, no. 4(1999): 387–91.

Mitchell, E. D. "An ESP Test from Apollo 14." *Journal of Parapsychology* 35, no. 2 (1971): 89–107.

Morehouse, David. *Psychic Warrior: Inside the CIA's Stargate program.* New York: St. Martin's Press, 1996.

Nelson, R. D., D. I. Radin, R. Shoup, and P. Bancel. "Correlation of Continuous Random Data with Major World Events." *Foundations of Physics Letters* 15, no. 6 (2002): 537–50.

Oteri, Laura. *Quantum Physics and Parapsychology: Proceedings of an International Conference Held in Geneva, Switzerland, August 26–27, 1974.* New York: Parapsychology Foundation, 1975.

"Outline of CIA Project on ESP." 1952 memorandum retrieved under Freedom of Information Act. In Martin Ebon, *Psychic Warfare: Threat or Illusion?* New York: McGraw-Hill, 1983.

Pagels, Heinz R. *The Cosmic Code: Quantum Physics as the Language of Nature.* New York: Simon and Schuster, 1982.

Palladino, Paola, and Rossana De Beni. "When Mental Images Are Very Detailed: Image Generation and Memory Performance as a Function of Age." *Acta Psychologica* 113, no. 3 (2003): 297–314.

Palmer, John A., Charles Honorton, and Jessica Utts. "Reply to the National Research Council Study on Parapsychology." Research Triangle Park, NC: Parapsychology Association, 1988.

Parapsychology Debate and Open Peer Commentary. In *Behavioral and Brain Sciences* 10, no. 4 (1987).

Parker, A., and G. Brusewitz. "A Compendium of the Evidence for Psi." *European Journal of Parapsychology* 18(2003): 33–51.

Peoc'h, Rene. "Chicken Imprinting and the Tychoscope: An ANPSI Experiment." *Journal of the Society for Psychical Research* 55, no. 810 (1988): 1–9.

————. "Psychokinetic Action Of Young Chicks on an Illuminated Source." *Journal of Scientific Exploration* 9 (1988): 223–29.

Puthoff, Harold E. "CIA-Initiated Remote Viewing at Stanford Research Institute." Available at http://www.biomindsuperpowers. com/Pages/CIA-InitiatedRV.html.

Radin, Dean I. *The Conscious Universe: The Scientific Truth of Psychic Phenomena.* New York: HarperEdge, 1997.

————. *Entangled Minds: Extrasensory Experiences in a Quantum Reality.* New York: Paraview Pocket Books, 2006.

Radin, Dean, Roger Nelson, York Dobyns, and Joop Houtkooper. "Reexamining Psychokinesis: Comment on Bösch, Steinkamp, and Boller." *Psychological Bulletin* 132, no. 4 (2006): 529–32.

Radin, D. I., and M. J. Schlitz. "Gut Feelings, Intuition, and Emotions: An Exploratory Study." *Journal of Alternative and Complementary Medicine* 11, no. 1 (2005): 85–91.

Randi, James. *The Magic of Uri Geller: By the Amazing Randi.* New York: Ballantine, 1975.

"Report of a Workshop on Experimental Parapsychology." International Security and Commerce Program, Office of Technology Assessment, United States Congress, February 22, 1989.

Rogo, D. Scott. *Miracles: A Parascientific Inquiry into Wondrous Phenomena.* New York: Dial Press, 1982.

Ronson, Jon. *The Men Who Stare at Goats.* New York: Simon and Schuster, 2006.

"Row Over Sargent's Psi Experiments Erupts with Evidence of Carelessness and Cheating." *Skeptical Inquirer* XII, no. 3 (Spring 1988): 226–31.

Schlitz, M., D. I. Radin, B. F. Malle, S. Schmidt, J. Utts, and G. L. Yount. "Distant Healing Intention: Definitions and Evolving Guidelines for Laboratory Studies." *Alternative Therapies in Health and Medicine* 9, no. 3 (May/June 2003): supplement pages 31–43.

Schmeidler, Gertrude Raffel, and R. A. McDonnell. *ESP and Personality Patterns*. New Haven, CT: Yale University Press, 1958.

Schmidt, Helmut J. "A Logically Consistent Model of a World with Psi Interaction." In Oteri, *Quantum Physics and Parapsychology*, 205–28.

———. "PK Effect on Pre-Recorded Targets." *Journal of the American Society for Psychical Research* 70 (July 1976): 267–91.

———. "Collapse of the State Vector and Psychokinetic Effect." *Foundations of Physics* 12, no. 6 (1982): 565–81.

———. "Comparison of a Teleological Model with a Quantum Collapse Model of Psi." *Journal of Parapsychology* 48, no. 4 (1984): 261–76.

Schnabel, Jim. *Remote Viewers: The Secret History of America's Psychic Spies*. New York: Dell, 1997.

———. "A Proposed Measure for Psi-Induced Bunching of Randomly Spaced Events." *Journal of Parapsychology* 64 (September 2000): 300–16.

Schwartz, Gary E., with William L. Simon. *The Truth about Medium: Extraordinary Experiments with the Real Allison DuBois of NBC's Medium and Other Remarkable Psychics*. Charlottesville, VA: Hampton Roads, 2005.

Schwartz, Stephan A. *The Alexandria Project: The Engineering of Psi, Volume Two*. Lincoln, NE: Authors Guild, 2001 (reprint with new appendix of Delacourte Press edition, 1983).

———. *Opening to the Infinite: The Art and Science of Nonlocal Awareness*. Austin, TX: Nemoseen, 2007.

Sheldrake, Rupert. *Dogs That Know When Their Owners Are Coming Home: And Other Unexplained Powers of Animals.* New York: Crown, 1999.

————. *The Sense of Being Stared At: And Other Aspects of the Extended Mind.* New York: Crown, 2003.

Shermer, Michael. *The Borderlands of Science: Where Sense Meets Nonsense.* Oxford: Oxford University Press, 2001.

Silverberg, Robert. *Dying Inside.* New York: Scribner, 1972.

————. *The Stochastic Man.* New York: Harper and Row, 1975.

Sinclair, Upton. *Mental Radio: Does it work and how?* London: T. Werner Laurie, 1930.

Stanford, Rex G. "An Experimentally Testable Model for Spontaneous Psi Events: A Review of Related Evidence and Concepts from Parapsychology and Other Sciences." In Krippner, ed., *Advances in Parapsychological Research,* vol. 6.

Steinkamp, Fiona. "Forced-Choice Experiments: Their Past and Their Future." In Michael Thalbourne and Lance Storm, eds., *Parapsychology in the Twenty-First Century: Essays on the Future of Psychical Research.* Jefferson, NC: McFarland, 2005.

Stokes, Douglas M. "Theoretical Parapsychology." In Krippner, ed., *Advances in Parapsychological Research.* vol. 5.

————. "The Shrinking Filedrawer: On the Validity of Statistical Meta-analyses in Parapsychology." *Skeptical Inquirer* 25, no. 3 (2001): 22–25.

————. "A History of the Relationship between Statistics and Parapsychology." *Journal of the American Society for Psychical Research* 96 (2002): 15–73.

Storm, L., and S. Ertel. "Does Psi Exist? Comments on Milton and Wiseman's (1999) Meta-Analysis of Ganzfeld Research." *Psychological Bulletin* 127, no. 3 (2001): 424–33

Targ, Russell, and Keith Harary. *The Mind Race: Understanding and Using Psychic Abilities.* New York: Villard Books, 1984.

Targ, Russell, and Jane Katra. *Miracles of Mind: Exploring Nonlocal Consciousness and Spiritual Healing.* Novato, CA: New World Library, 1998.

Targ, Russell, and Harold Puthoff. "Information transmission under conditions of sensory shielding." *Nature,* 1974, 251. 602–607.

————."A perceptual channel for information transfer over kilometer distances: Historical perspective and recent research." *proceedings of the IEEE,* 1976, 64. 329–354.

————. *Mind-Reach: Scientists Look at Psychic Ability.* London: Jonathan Cape, 1977.

Tart, Charles T. *PSI: Scientific Studies of the Psychic Realm.* New York: Dutton, 1977.

————. "Life in the World Simulator: Altered States, Identification, Multiple Personality, and Enlightenment." Address given at the Seventh International Conference of Multiple Personality/Dissociative States, Chicago, November 7–11, 1990.

————. "Multiple Personality, Altered States and Virtual Reality: The World Simulation Process Approach." *Dissociation* 3 (1991): 222–33. Available at www.paradigm-sys.com/cttart/.

Taylor, John. *Superminds: An Enquiry into the Paranormal.* London: Macmillan, 1975.

————. *Science and the Supernatural: An Investigation of Paranormal Phenomena.* London: Granada, 1980.

Taylor, Robin. "Evolutionary Theory and Psi: Reviewing and Revising Some Need-Serving Models in Psychic Functioning." *Journal of the Society for Psychical Research* 67 (January 2003).

Thouless, Robert H. *From Anecdote to Experiment in Psychical Research.* London: Routledge and Kegan Paul, 1972.

Todeschi, Kevin J. "RV describes Saddam's capture before it occurs." Virginia Beach, VA: *Venture Inward,* March/April, 2004: 30–31.

Utts, Jessica. "Reponse to Ray Hyman's Report of September 11, 1995. Evaluation of Program on Anomalous Mental Phenomena." Available at http://anson.ucdavis.edu/~utts/response.html.

Von Lucadou, W. "The Model of Pragmatic Information (MPI)," *European Journal of Parapsychology* 11 (1991): 58–75.

———. "Hans in Luck: The Currency of Evidence in Parapsychology." *Journal of Parapsychology* 65 (2001): 3–16.

Walker, Evan Harris. "Foundations of Paraphysical and Parapsychological Phenomena." In Oteri, *Quantum Physics and Parapsychology,* 1–53.

———. "A Review of Criticisms of the Quantum Mechanical Theory of Psi Phenomena." *Journal of Parapsychology* 48, no. 4 (1984): 277–332.

———. *The Physics of Consciousness: Quantum Minds and the Meaning of Life.* Cambridge, MA: Perseus Books, 2000.

Webb, James. *The Occult Establishment.* La Salle, IL: Open Court Publishing, 1976.

Wendell, John P. "More on Jahn's Statistics." *Skeptical Inquirer* 16, no. 1 (1991): 89–90.

Wheeler, John Archibald. *At Home in the Universe.* New York: American Institute of Physics, 1994.

Wolman, Benjamin B., ed. *Handbook of Parapsychology.* New York: Van Nostrand Reinhold, 1977.

INDEX

A

AAAS symposium on reverse causality, 190

active scanning, 241

adaptation, 42, 221, 232, 242

Afshar, Shahriar, 189

afterlife, 71, 303–04

Aickin, Mikel, 154

AIR (American Institutes for Research), 77, 124–26, 133

Akashic records, 172

Alcock, James, 308

alleles, 232–33

Amelie, 161, 249, 251–52

Anderson, Jack, 54

animist projection, 277

"Anomalies: Analysis and Aesthetics" (Jahn), 33

anomalous cognition (AC), 4–5, 77, 123, 126–29, 138, 141, 261

anomalous perturbation (AP), 4, 126, 213, 261, 285

Arquette, Patricia, 40

Association for Research and Enlightenment (ARE), 95, 101, 103

Astounding Science Fiction magazine, 55, 219, 294

astrology, 18, 55, 135–36

Atkins, Peter, 165–66

Atlantic University, 94–95, 102–03

Atmanspacher, Harald, 122, 265

Australian aborigines, 223

B

"Ballad of Bowsprit Bear's Stead, The," (Broderick), 235

Bannister, Roger, 109–10, 113, 120, 132

baseline (BL), 24, 28, 34, 119, 283

BBC, 166

beacon (traveling RV agent), 29, 59, 83, 111–12

bean-counters, 10, 16, 288

Beauregard, Mario, 261

Bell, John, 181–82

Benveniste, Jacques, 183

Bester, Alfred, 230

Bieler, Peter, 92

Bierman, Dick J., 122, 143, 151–53, 162–63, 202–03, 208, 218, 269

bin Laden, Osama, 99, 100–01

binary descriptors, 30

bioentanglement, 258

biological-psychological virtual reality (BPVR), 302

black holism, 181

Blake, William, 173

blindsight, 216–19

Bohr, Niels, 20, 177, 182–83

boredom, 162, 242

Borges, Jorge Luis, 302

Born, Max, 279

Bradbury, Ray, 192, 292

Brian, Denis, 11 *et seq.*

British Association for the Advancement of Science (the BA), 165